FUNDAMENTALS OF MODERN STATISTICAL METHODS

Springer
New York
Berlin
Heidelberg
Hong Kong
London
Milan
Paris
Tokyo

RAND R. WILCOX

FUNDAMENTALS OF MODERN STATISTICAL METHODS

SUBSTANTIALLY IMPROVING
POWER AND ACCURACY

 Springer

Rand R. Wilcox
Department of Psychology
University of Southern California
Los Angeles, CA 90089-1061
USA
rwilcox@wilcox.usc.edu

Library of Congress Cataloging-in-Publication Data
Wilcox, Rand R.
   Fundamentals of modern statistical methods : substantially improving
power and accuracy / Rand R. Wilcox
     p. cm.
   Includes bibliographical references and index.
   ISBN 0-387-95157-1 (hardcover : alk. paper)
   QA276.W48   2001
   519.5—dc21   00-047082

Printed on acid-free paper.

© 2001 Springer-Verlag New York, Inc.
All rights reserved. This work may not be translated or copied in whole or in part without the written permission of the publisher (Springer-Verlag New York, Inc., 175 Fifth Avenue, New York, NY 10010, USA), except for brief excerpts in connection with reviews or scholarly analysis. Use in connection with any form of information storage and retrieval, electronic adaptation, computer software, or by similar or dissimilar methodology now known or hereafter developed is forbidden. The use of general descriptive names, trade names, trademarks, etc., in this publication, even if the former are not especially identified, is not to be taken as a sign that such names, as understood by the Trade Marks and Merchandise Marks Act, may accordingly be used freely by anyone.

Printed in the United States of America.   (MVY)

9 8 7 6 5 4 3 2

ISBN 0-387-95157-1   SPIN 10980224

Springer-Verlag is a part of *Springer Science+Business Media*

springeronline.com

To Bryce, Quinn, and Karen

If at first the idea is not absurd, then there is not hope for it.
— ALBERT EINSTEIN

Everyone believes in the [normal] law of errors, the experimenters because they think it is a mathematical theorem, the mathematicians because they think it is an experimental fact.
— HENRI POINCARÉ

Each generation that discovers something from its experience must pass that on, but it must pass that on with a delicate balance of respect and disrespect, so that the race ... does not inflict its errors too rigidly on its youth, but it does pass on the accumulated wisdom plus the wisdom that it may not be wisdom.
— RICHARD FEYNMAN

PREFACE

This book is about understanding basic statistics from the point of view of modern developments and insights achieved during the last forty years. The strength and utility of classical statistical methods is a marvel to behold. They have benefited our lives in countless ways. Yet, about two hundred years ago, there were already signs that many conventional methods used today contain fundamental problems that are relevant in many applied settings. In hindsight it is easy to see why these fundamental problems could not be addressed until fairly recently. Indeed, it took three major developments to bring about practical solutions to the problems described in this book: Better theoretical tools for understanding and studying non-normality, more effective methods for making inferences about populations based on a random sample of observations, and fast computers. Without these developments, some of the insights and advances from the first half of the nineteenth century were doomed to remain nothing more than mathematical curiosities. Applied reseachers would never bother about these results because they yielded methods that are computationally intractable, and there was weak evidence that they could be safely ignored. But during the latter half of the twentieth century, things began to change dramatically. Computers made a host of new statistical methods a practical reality. With the advent of new mathematical methods for understanding how non-normality affects more traditional techniques, the need for better methods—methods that deal effectively with non-normality—became evident.

The technological advances of the last forty years have made it possible to get more accurate and more revealing information about how groups of individuals differ and how variables are related. Unfortunately, standard training in basic statistics does not prepare the student for understanding the practical problems with conventional techniques or why more modern tools might offer a practical advantage. Indeed, at first glance it might seem that modern methods could not possibly have any practical value, yet the increased accuracy they provide can be substantial in commonly occurring situations. The result is an ever-increasing gap between state-of-the-art methods versus techniques commonly used. The goal in this book is to help bridge this gap.

Part I of this book is aimed at providing a verbal and graphical explanation of why standard methods can be highly misleading. Technical details are kept to a minimum. The hope is to provide understanding with little or no mathematics. Another goal in Part I is to provide a framework for intuitively understanding the practical advantages of modern techniques. Presumably portions of Part I cover basic concepts already familiar to most readers. However, various perspectives are not typically covered in an applied course. Readers with strong training in mathematical statistics can skim much of the material in Part I and then read Part II. Part II describes the most basic methods for dealing with the problems described in Part I. The list of modern methods covered here is far from exhaustive. A more advanced book is needed to cover many of the complex problems that arise, yet some of the methods covered here are at the cutting edge of technology. The use of these methods is supported by many journal articles published by a wide range of statisticians. In fact there is a strong mathematical foundation for these techniques, but this topic goes well beyond the scope of this book. The goal here is to address what has become a more pressing issue: Explaining modern methods to applied researchers who might benefit from their use.

I would like to thank John Kimmel for his support and interest in this book. Without his insight and encouragement, this book would not exist. I would also like to thank two anonymous referees for many helpful and constructive suggestions.

Rand R. Wilcox
Los Angeles, CA

CONTENTS

PREFACE   vii

1 • INTRODUCTION

A Brief History of the Normal Curve   3
Empirical Studies Regarding Normality   5
Inferential Methods   7

PART I

2 • GETTING STARTED

Probability Curves   12
The Mean   14
The Median   17
A Weighted Mean   19

Variance	20
Measuring Error	22
Fitting a Straight Line to Data	24
A Summary of Key Points	30

3 • THE NORMAL CURVE AND OUTLIER DETECTION

The Normal Curve	32
Detecting Outliers	33
The Central Limit Theorem	38
Normality and the Median	44
A Summary of Key Points	47

4 • ACCURACY AND INFERENCE

Some Optimal Properties of the Mean	50
The Median Versus the Mean	52
Regression	55
Confidence Intervals	58
Confidence Interval for the Population Mean	60
Confidence Interval for the Slope	62
A Summary of Key Points	65

5 • HYPOTHESIS TESTING AND SMALL SAMPLE SIZES

Hypothesis Testing	68
The One-Sample T Test	72
Some Practical Problems with Student's T	78
The Two-Sample Case	82
The Good News About Student's T	84
The Bad News about Student's T	85
What Does Rejecting with Student's T Tell Us?	87
Comparing Multiple Groups	89
A Summary of Key Points	89
Bibliographic Notes	90

6 • THE BOOTSTRAP

Two Bootstrap Methods for Means	94
Testing Hypotheses	104
Why Does the Percentile t Bootstrap Beat Student's T	104
Comparing Two Independent Groups	105
Regression	107
Correlation and Tests of Independence	109
A Summary of Key Points	115
Bibliographic Notes	115

7 • A FUNDAMENTAL PROBLEM

Power	120
Another Look at Accuracy	123
The Graphical Interpretation of Variance	124
Outlier Detection	126
Measuring Effect Size	127
How Extreme Can the Mean Be?	127
Regression	129
Pearson's Correlation	132
More About Outlier Detection	134
A Summary of Key Points	134
Bibliographic Notes	135

PART II

8 • ROBUST MEASURES OF LOCATION

The Trimmed Mean	141
The Population Trimmed Mean	147
M-Estimators	149
Computing a One-Step M-Estimator of Location	153
A Summary of Key Points	157
Bibliographic Notes	157

9 • INFERENCES ABOUT ROBUST MEASURES OF LOCATION

Estimating the Variance of the Trimmed Mean	159
Inferences About the Population Trimmed Mean	166
The Relative Merits of Using a Trimmed Mean Versus a Mean	169
The Two-Sample Case	170
Power Using Trimmed Means Versus Means	172
Inferences about M-Estimators	174
The Two-Sample Case Using an M-Estimator	175
Some Remaining Issues	175
A Summary of Key Points	177
Bibliographic Notes	178

10 • MEASURES OF ASSOCIATION

What Does Pearson's Correlation Tell Us?	179
A Comment on Curvature	183
Other Ways Pearson's Correlation Is Used	184
The Winsorized Correlation	187
Spearman's Rho	190
Kendall's Tau	191
Methods Related to M-Estimators	193
A Possible Problem	193
Global Measures of Association	196
More Comments on Curvature	200
A Summary of Key Points	202
Bibliographic Notes	203

11 • ROBUST REGRESSION

Theil-Sen Estimator	207
Regression via Robust Correlation and Variances	210
L_1 Regression	211
Least Trimmed Squares	212
Least Trimmed Absolute Value	216
Least Median of Squares	217
Regression Outliers and Leverage Points	217
M-Estimators	220
The Deepest Regression Line	223

Relative Merits and Extensions to Multiple Predictors	223
Correlation Coefficients Based on Regression Estimators	224
A Summary of Key Points	225
Bibliographic Notes	227

12 • ALTERNATE STRATEGIES

Ranked-Based Methods for Comparing Two Groups	229
Permutation Tests	235
Extension to Other Experimental Designs	236
Regression Based on Ranked Residuals	240
Software	241
A Summary of Key Points	242
Bibliographic Notes	243

APPENDIX A	245
REFERENCES	249
INDEX	253

CHAPTER *1*

INTRODUCTION

If we measure the worth of an equation by how many disciplines use it, few can rival the equation for the normal curve. It plays a fundamental role in physics and astronomy as well as in manufacturing, economics, meteorology, medicine, biology, agriculture, sociology, geodesy, anthropology, communications, accounting, education, and psychology. The normal curve suggests a strategy toward a host of applied problems of great importance, and it even influences how many of us view the world in which we live. We have all encountered, for example, the notion that IQ scores follow a normal curve, or that this curve should be used to assign grades in school. The utility of the equation is not in doubt—it provides a useful solution to a wide range of problems. But our understanding of this curve—how it might mislead us in our attempts to model reality—has grown tremendously during the last forty years. As pointed out in hundreds of journal articles, for many applied problems the use of the normal curve can be disastrous. Even under *arbitrarily small* departures from normality, important discoveries are lost by assuming that observations follow a normal curve. These lost discoveries include both the detection of differences between groups of subjects and important associations among variables of interest. Even if differences are detected, the magnitude of these differences can also be grossly underestimated using a commonly employed strategy based on the normal curve, and the characterization of differences can be highly misleading. Associations among variables can also be grossly misunderstood. Moreover, some commonly recommended methods for dealing with nonnormality have been found to be completely useless. In some cases the normal curve even leads to the wrong answer no matter how many observations we might have.

The normal curve is one of the main characters in the story that follows, but perhaps of equal importance is the story of conventional hypothesis testing methods covered in a basic statistics course. It seems fair to say that according to conventional wisdom, these methods provide accurate results in virtually all situations that arise in practice. Some introductory books hint that practical problems might arise, but no details are given as to when and why the applied researcher should be concerned. Based on some results discussed in Section 3.3 of Chapter 3, some textbooks speculated that if there are at least twenty-five observations, standard methods perform quite well. But we now know that when using some commonly employed techniques, hundreds of observations might be necessary, and in some cases we get inaccurate results no matter how many observations we might have!

The good news is that practical methods for dealing with these problems have been developed. These modern techniques stem from three major developments: the theory of robustness that emerged in the 1960s; new inferential methods developed during the 1970s that beat our reliance on the so-called central limit theorem; and fast computers. These new methods have a strong theoretical foundation, and numerous simulation studies have shown that in many realistic situations they offer a tremendous advantage over more traditional techniques. Moreover, even when conventional assumptions are true, modern methods compete relatively well with standard procedures. But two important goals remain. The first is fostering an appreciation and understanding among applied researchers of the practical problems with traditional techniques. Why did a consensus emerge that traditional methods are insensitive to violations of assumptions, and why did it take so long to discover their problems? The second goal is to provide some sense of why modern methods work. They are not intuitive based on the training most of us receive. Indeed, at first glance it might seem that they could not possibly have any value. Yet, in light of results derived nearly two hundred years ago, some modern insights are not completely surprising.

A BRIEF HISTORY OF THE NORMAL CURVE

To understand and appreciate conventional wisdom regarding standard statistical techniques, it helps to begin with a brief history of the normal curve. The derivation of the normal curve is due to Abraham de Moivre and arose in the context of what we now call the binomial probability function. An ABC news program (20/20), doing a story on people undergoing surgery, provides a modern example of why the binomial probability function is important. A serious problem is that some patients wake up during surgery—they regain consciousness and become aware of what is being done to them. These poor individuals not only have a conscious experience of their horrific ordeal, they suffer from nightmares later. Of course, patients are given medication to render them unconscious, but the amount of a drug to be administered is determined by body weight, which results in some patients waking up. To deal with the problem, some doctors tried monitoring brain function. If a patient showed signs of regaining consciousness, more medication was given to keep them under. In the actual news story, two hundred thousand patients underwent the new method and zero woke up, so the probability of waking up was estimated to be zero. But hospital administrators, concerned about the added cost of monitoring brain function, argued that two hundred thousand was too small a sample to be reasonably certain about the actual probability. How many times should the new method be used to get an accurate estimate of the probability that someone will regain consciousness?

This type of problem, where the goal is to determine the precision of the estimated probability of success, appears to have been first posed by Jacob Bernoulli about three hundred years ago. Initially de Moivre felt he could make no contribution, but despite his reluctance he made one of the great discoveries of all time. The specific problem he had in mind was flipping a coin one thousand times, with the probability of a head equal to .5 on each flip. He wanted to determine, for example, the probability that the number of heads would be between 450 and 550 (in which case the observed proportion of heads would be between .45 and .55). The exact probability was too unwieldy to compute, so de Moivre set out to find an approximate but reasonably accurate solution. He worked on this problem over a twelve-year period, which culminated in the equation for the normal curve in 1733.

From an applied point of view, de Moivre's equation generated no immediate interest. The normal curve provides a basis for dealing with the issue raised by Bernoulli, but it would be years before a practical solution, based on the normal curve, would be derived. It was the combined work of Laplace and Gauss that helped to catapult the normal curve to prominence in statistics and the science of uncertainty. Contrary to what might be thought, it was mathematical expediency that first initiated interest in the normal curve versus any empirical investigations that it might have practical value. Indeed, Laplace made attempts at using other curves for modeling observations, but they proved to be mathematically intractable.

In 1809, Gauss gave an argument for the normal curve that went like this. First, he assumed that if we were able to obtain a large number of observations, a plot of the observations would be symmetric about some unknown point. To elaborate in more concrete terms, imagine we want to determine the time it takes light to travel between two points. For illustrative purposes, assume that, unknown to us, the exact time is ten seconds, but due to observational error the exact time cannot be determined. Further assume that observed values are distributed symmetrically around the exact time. So the probability of getting a value of 10.1 or larger is the same as the probability of getting a value of 9.9 or smaller. Similarly, a value of 10.5 or larger has the same probability as a value of 9.5 or smaller. More generally, for any constant c we might pick, the probability of getting a value of $10 + c$ or greater is the same as the probability of getting a value of $10 - c$ or less. Now imagine we make six measurements and get the values 10.1, 8, 8.9, 9.7, 11, and 10.3. How can we combine these values to get an estimate of the exact time? One possibility is to simply compute the mean, the average of the six values, which yields 9.67. Another possibility is to use the median of these six values, which is the average of the two middle values when they are put in ascending order. Here the median is 9.9. Both estimates are in error—they differ from the true time, 10. Gauss assumed that among the methods one might use to estimate the true time, generally the mean would be more accurate than the median or any other method one might use to combine the six values. He then showed that by implication, the observed measurements arise from a normal curve. One can turn this argument around. If we assume that observations are normally distributed and centered around the true time, then the optimal estimate of the true time is the mean.

As noted by the prominent statistician and historian Stephen Stigler, Gauss's argument is far from compelling—it is both circular and a non sequitur. There is no reason to assume that the mean is the optimal method for combining observations. Indeed, by the year 1818, Laplace was aware of situations where the median beats the mean in accuracy. And as early as 1775, Laplace found situations where the mean was not optimal. Simultaneously, Gauss had no empirical reason to assume that observations follow a normal curve. Of course, Gauss was aware that his line of reasoning was less than satisfactory, and he returned to this issue at various points during his career. His efforts led to the so-called Gauss-Markov theorem, which is described in Chapter 4.

Why did Gauss assume that a plot of many observations would be symmetric around some point? Again the answer does not stem from any empirical argument, but rather a convenient assumption that was in vogue at the time. This assumption can be traced back to the first half of the eighteenth century and is due to Thomas Simpson. Circa 1755, Thomas Bayes argued that there was no particular reason for assuming symmetry; Simpson recognized and acknowledged the merit of Bayes's argument, but it was unclear how to make any mathematical progress if asymmetry was allowed.

The conventional method of justifying the normal curve is to appeal to Laplace's central limit theorem, which he publicly announced in 1810. (The word *central* is intended to mean fundamental.) A special case of this theorem describes conditions under which normality can be assumed to give accurate results when working with means. The simple interpretation of this theorem is that inferences based on means can be made with the normal curve if one has a reasonably large number of observations. However, serious practical problems remain. One is getting some sense of what constitutes a reasonably large number of observations. Views about this issue have changed substantially in recent years. A second concern has to do with variances, but the details are too involved to discuss now. Again very serious problems arise, the details of which will be explained in due course, particularly in Chapter 7. A third concern is that regardless of how many observations one might have, certain practical problems persist.

Yet another path to the normal curve is the so-called least squares principle which is discussed in some detail in this book. For now, suffice it to say that least squares does not provide a satisfactory justification for the normal curve, and in fact it reflects serious practical problems associated with the normal curve, as we shall see. But from a mathematical point of view it is extremely convenient, and Gauss exploited this convenience in a very impressive fashion which no doubt played a major role in the adoption of the normal curve.

EMPIRICAL STUDIES REGARDING NORMALITY

In 1818, Bessel conducted the first empirical investigation that focused on whether observations follow a normal curve. The data dealt with the declination and ascension of some stars. He concluded that the normal curve provides a good approximation, but he noted that there seemed to be more extreme observations than predicted. One problem Bessel faced was understanding what type of departure from normality might cause practical problems. It appears he might have witnessed a departure from normality that was a serious concern, but he did not have the tools and framework for addressing this problem.

Subsequent studies found that observations do not always follow a normal curve, but at least some prominent scientists of the nineteenth century chose to believe that it was at the heart of phenomena that interested them. Why? One explanation stems from the goal of finding some deterministic model that explains all observations. Adolphe Quetelet is an excellent early example. Laplace's central limit theorem generated a sense of wonder in Quetelet that out of chaos comes order. But Quetelet's faith in the normal curve was due to reading too much into Laplace's result. Indeed, Laplace himself did not conclude that normality should be assumed, and he had theoretical reasons for being concerned about one implication of the normal curve, which will be discussed in Chapter 3.

During the early portion of his career, Karl Pearson, one of the most prominent individuals in statistics, had a firm belief in the bell (normal) curve derived by de Moivre. So strong was his belief that in his first paper on the science of generalization and measurement, he dubbed the bell curve the "normal" curve. That is, the equation giving the bell curve was thought to be the natural curve—the curve we should expect. Pearson wrote that if we take many measurements and we get a normal curve, we have what he characterized as a "stable condition; there is production and destruction impartially around the mean."

To his credit, Pearson conducted empirical studies to see whether his belief could be verified. One implication of the normal curve is that observations should be symmetrically distributed around some central value, but Pearson found that this was often not the case. At one time he thought that these nonnormal distributions were actually mixtures of normal distributions and he proposed that efforts be made to find techniques for separating these nonnormal distributions into what he presumed were the normal components. Eventually, however, he abandoned this idea and invented a system of curves for approximating what we find in nature. Unfortunately, Pearson's system of curves does not deal with many of the practical problems that plague methods based on the normal curve.

In more recent years, new empirical studies have been published that raise further concerns about assuming normality. Even in situations where nonnormality was thought to pose no problems, modern theoretical insights reveal that this is not always the case. For example, it might be thought that if a plot of observations appears to be reasonably symmetric, normality can be assumed, but often the exact opposite is true. In fact, even if observations are perfectly symmetric about some central value, it can be highly advantageous to abandon the assumption of normality.

INFERENTIAL METHODS

A fundamental goal in statistics is making inferences about a large population of individuals based on only a subset of individuals available to us. In the example about patients waking up during surgery, there is interest in knowing what the sample of two hundred thousand patients undergoing the new treatment tells us about the millions of individuals who will undergo surgery in the future. In 1811, Pierre-Simon Laplace developed a strategy for making inferences (generalizations from a sample of observations to some population of interest) that dominates in applied work today. Prior to 1811, the only available framework for making inferences was the so-called method of inverse probability, what we now call a Bayesian method. Laplace developed a new strategy based on what is called the frequentist point of view. In effect, Laplace created a controversy that is still with us today: How and when should a Bayesian point of view be employed? Ironically, even though the Reverend Thomas Bayes first suggested the notion of inverse probability, it was Laplace who championed the Bayesian view and made major contributions to its use. Indeed, it appears that Laplace developed this approach independently of Bayes, yet at age 62 he created a new approach that would dominate in applied work two hundred years into the future.

In 1814 Laplace used his frequentist view to develop a new approach to computing what is called a confidence interval. The exact meaning of a confidence interval is described in Chapter 4. For the moment it suffices to know that confidence intervals are intended to reflect how well we can generalize to a population of individuals under study. In the surgery example, the proportion of individuals waking up was zero. Based on this information, is it reasonable to rule out the possibility that for future patients, the probability of waking up is less than .2 or less than .1? Confidence intervals represent an approach to this problem. Today, Laplace's method for computing a confidence interval is taught in every introductory statistics course. The great mathematician Carl Gauss quickly endorsed Laplace's method and made very important contributions to it (particularly in the context of what we now call the least squares principle), but it was still slow to catch on. Major refinements and extensions were developed during the first half of the twentieth century, including Sir Ronald Fisher's methods for making inferences about means, and the Neyman-Pearson framework for hypothesis testing. Like Laplace's technique, these newer methods are based on the assumption that observations have a normal distribution. Laplace realized that there is no particular reason for assuming normality, and his strategy for dealing with this issue was to appeal to his central limit theorem which is described in Chapter 3. In effect, when computing a confidence interval, he found a method that provides reasonably accurate results under random sampling provided that the number of observations is sufficiently large. For a wide range of problems an additional assumption is typically made, what we now call homogeneity of variance, a notion that will be discussed in detail in Chapter 4. Violating this assumption was once thought to have no practical implications, but as we shall see, many recent journal articles report that serious problems do indeed arise.

From a practical point of view, when invoking the central limit theorem, we have two rather obvious concerns. The first, which was already mentioned, is determining what constitutes a reasonably large sample size. A second problem is assessing the practical consequences of having unequal variances. A third concern is whether there are any unforeseen problems with using means in particular and the least squares principle in general. Recent insights about the first two issues are described in Chapters 2 through 5 and 7. As for the third issue, it turns out that unforeseen problems do indeed arise, as we shall see throughout this book.

PART ONE

CHAPTER *2*

GETTING STARTED

The goals in Part I of this book are to describe and explain basic methods typically covered in an applied statistics course, but from a perspective that helps the reader appreciate, understand, and conceptualize the practical problems with standard statistical techniques. Another goal is to provide a foundation for understanding why some modern methods offer a distinct advantage in applied work. Yet another important goal is understanding why certain common strategies for dealing with nonnormality fail. Perhaps the most striking problem with standard statistical methods is described in Chapter 7, but very serious problems are also covered in Chapters 3, 4, and 5. This chapter covers some basics about measures of location and scale, but even here some important concepts and perspectives are introduced that are not typically covered in an introductory course. So even if the reader has had an elementary statistics course, it is strongly recommended that the first part of this book be read carefully.

PROBABILITY CURVES

It is assumed that the reader is familiar with basic probability, but to make this book as self-contained as possible, a quick review of graphical representations of probabilities is given here.

Two graphical methods are typically used to represent probabilities. The first applies to discrete variables where probabilities are represented by the height of spikes. Imagine, for example, we flip a coin having probability .5 of producing a head. If we flip the coin ten times and count the number of heads, the probabilities are as shown in Figure 2.1 (assuming the outcomes of each flip are independent and the probability of a head remains .5 for each flip).

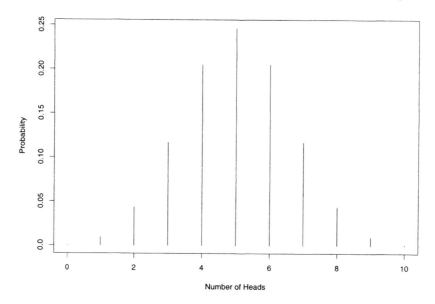

FIGURE 2.1 • For discrete variables, probabilities are graphically represented by the height of spikes. In this particular case, the height of each spike represents the probability of the corresponding number of heads when a coin is flipped ten times. For example, the height of the middle spike is .246 and represents the probability of getting exactly five heads.

For continuous variables (variables that can take on any value over some specified interval), probabilities are represented by the area under a curve called a *probability density function*. One of the earliest families of probability curves is due to Laplace; an example of it appears in Figure 2.2. Although not indicated in Figure 2.2, this particular probability curve extends over all real numbers. (It is defined for all numbers between minus infinity and infinity.) The area of the shaded region in Figure 2.2 represents the probability that an observation has a value between .5 and 1.

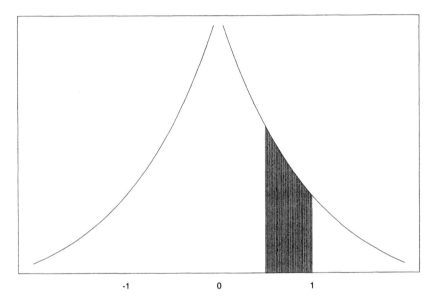

FIGURE 2.2 • For continuous variables, probabilities are represented graphically by the area under a curve called a probability density function. The curve used here belongs to a family of probability curves known as Laplace distributions. The area of the shaded region is the probability that an observation is between .5 and 1.

A requirement of all probability density functions is that the area under the curve is one. The area cannot be greater than one because this would mean we have a probability greater than one, which is impossible, and it cannot be less than one because the curve is intended to represent the probability of all possible events. If, for example, we measure how much weight someone loses by going to Jenny Craig, we can be certain that this value is between minus infinity and infinity. That is, the probability is one.

There are many families of probability curves in addition to Laplace's family, including the family of normal probability curves, which will be formally introduced in Chapter 3.

THE MEAN

Next we consider how one might choose a single value to represent the typical individual or thing under study. Such values are called *measures of location* or *measures of central tendency*. For example, we might want to know the typical cholesterol level of an adult who smokes or the typical income of a real estate broker living in Los Angeles. A natural and common strategy is to use an average. If we could measure the cholesterol level of *all* adults who smoke and we averaged the results, we would know what is called the *population mean*, which is typically labeled μ. Rarely can all individuals be measured, so we use a random sample of individuals to estimate μ. For example, if we measure the cholesterol levels of six adults and get

$$200, 145, 320, 285, 360, 230,$$

the average of these six values is 256.7, and this is the usual estimate of μ (the population mean) based on the available data. The average of these six numbers is an example of a *sample mean*, which is typically labeled $\bar{X}$ and read as X bar. A fundamental issue is how well the sample mean $\bar{X}$ estimates the population mean μ, the average we would get if *all* individuals could be measured. We will discuss this problem in great detail, but an even more fundamental issue is whether we should be content using the population mean to reflect the typical individual under study.

If a probability curve is symmetric about some value, there is general agreement that the population mean represents a reasonable measure of the typical individual or measurement. One example is Laplace's probability curve shown in Figure 2.2. For such probability curves, the central value can be shown to correspond to the population mean. In Figure 2.2, the population mean is zero. Moreover, the probability that an observation is less than this central value is .5.

For a more concrete example, imagine you are interested in the diastolic blood pressure of adults living in New York City, and assume the population mean is 95. (The average diastolic blood pressure of all adults living in New York is $\mu = 95$.) Similar to the light example in Chapter 1, observations are said to be symmetric about 95 if for any constant c we might pick, the probability of getting a blood pressure reading greater than $95 + c$ is the same as the probability of getting a reading that is less than $95 - c$. So under symmetry around 95, the probability of getting a reading greater than 98 is the same as getting a reading less than 92, the probability of getting a reading greater than 105 is the same as getting a reading less than 85, and so on.

Exact symmetry implies that if we were to randomly sample an adult living in New York, the probability that her diastolic blood pressure is less than 95 is exactly .5. Rather than Laplace's distribution in Figure 2.2, the probability curve might be bell-shaped, as shown in Figure 2.3.

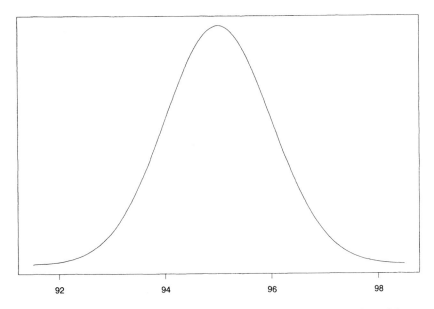

FIGURE 2.3 • Another example of a symmetric probability curve that might be used to represent probabilities associated with diastolic blood pressure. The population mean for a symmetric probability curve is the value around which the curve is centered. In this particular case, the curve is centered around 95. Again, the area under this curve is one, so the area under the curve and to the left of 95 is .5. That is, according to this probability curve, there is a .5 probability that a randomly sampled adult will have a diastolic blood pressure less than 95.

In actuality, probability curves are never exactly symmetric, but the assumption of symmetry seems to provide a reasonable approximation of reality in some situations. A practical issue, however, is whether situations ever arise where asymmetry raises doubts about using the population mean to represent the typical measurement, and the answer is an unequivocal yes.

Why is it that so many marriages in the United States end in divorce? One proposed explanation is that humans, especially men, seek multiple sexual partners, and that this propensity is rooted in our evolutionary past. In support of this view, some researchers have pointed out that when young males are asked how many sexual partners they desire over their life time, the average number has been found to be substantially higher than the responses given by females. However, other researchers have raised concerns—based on empirical results—about using a mean in such studies. In one such study by W. Pedersen, L. Miller, and their colleagues, one hundred five males were asked how many sexual partners they desired over the next thirty years. The average of the responses was found to be 64.3. The typical young male wants about 64 sexual partners? Looking a little more closely at the data, we see that five responded 0; they want no sexual partners over the next thirty years, and fifty of the one hundred five males gave a response of 1, by far the most common response among the males interviewed. Moreover, more than 97 percent of the responses are less than the average. Surely there is some doubt about using 64.3 as a reflection of the typical male.

What is going on? Looking at the data again, we see that one individual responded that he wanted 6,000 sexual partners over the next thirty years. If we remove this one value and average those that remain, we get 7.8. So, we see that a single individual can have a large impact on the average. However, there is concern about using 7.8 to represent the typical male because 78 percent of the observations are less than 7.8. Two males responded that they wanted 150 sexual partners, which again has a large influence on the average. If we label 7.8 as typical, this might conjure up the notion that most young males want about eight sexual partners over the next thirty years. But this is not what the data suggest because the majority of males want one partner or less.

The feature of the sample mean just illustrated is that its value is highly susceptible to extreme observations. Extreme values among a batch of numbers are called *outliers*. Among males interviewed about the number of sexual partners they desire over the next thirty years, 6,000 is an outlier; it is far removed from the bulk of the observed values. It is not remotely typical of the one hundred five males interviewed, and it offers an opportunity to be misleading about the sexual attitudes of most males.

One might argue that if more observations had been sampled, surely the mean would be closer to the center of the observations, but this is not necessarily so. Even if we sample infinitely many observations, in which case we would know the population mean exactly, it is possible for μ to be as far removed from the majority of values as you like. For example, there might be a .98 probability that an observation is less than the population mean. The details are a bit too involved to give now, but we will return to this topic later. However, the feasibility of this claim can be gleaned by asking the following question: Regardless of how many observations we might have, how many outliers does it take to make the sample mean arbitrarily large or small? The answer is 1. That is, a single observation can completely dominate the value of the sample mean.

It turns out to be convenient to express this last result in terms of what is called the *finite sample breakdown point* of the sample mean. If we have n values or observations, the smallest proportion of observations that can result in the sample mean being arbitrarily large or small is the finite sample breakdown point of the sample mean. This proportion is $1/n$. Notice that as n gets large, the finite sample breakdown point of the sample mean goes to zero. That is, an arbitrarily small subpopulation of individuals can cause the population mean to be arbitrarily small or large, regardless of what the bulk of the values happen to be.

THE MEDIAN

For a continuous probability curve, the *population median* is the number such that there is a .5 probability of an observation being less than it. If, for example, there is a .5 probability that an observation is less than 9, then 9 is the population median. In Figure 2.2 there is a .5 probability that an observed value will be less than 0 (the probability curve is symmetric about 0), so 0 is the median. For symmetric probability curves, the population mean and median are identical. In Figure 2.3, both the population mean and median are 95. (For discrete measures, there is a formal definition of the population median, but the details are not important here.)

Generally, the mean associated with an asymmetric probability curve differs from the median. In some cases the difference can be substantial, as illustrated in Figure 2.4. Note that the mean lies in the right tail and is relatively removed from the most likely values. In fact, for this particular curve, the probability that an observation is less than the mean (7.6), is approximately .74.

Like the population mean, the population median is not known in general and must be estimated based on observations we make. The most common method for estimating the population median is with the so-called *sample median*, which is computed as follows. If the number of observations is odd, the sample median is just the middle value after putting them in ascending order. For example, if we observe the values 21, 2, 36, 29, and 18, then putting them in ascending order yields

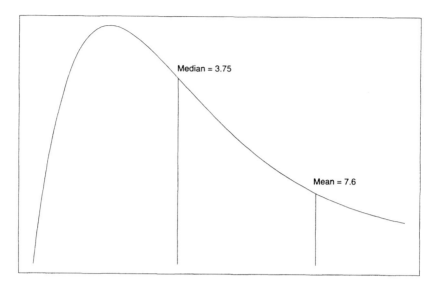

FIGURE 2.4 • An example of a probability curve where there is a rather substantial difference between the mean and median. The median, 3.75, is near the "center" of the curve in the sense that there is a .5 probability that a random sample observation will be less than 3.75 or greater than 3.75. In contrast, the mean is a more extreme value because there is about a .74 probability that a random sample observation will be less than the mean.

$$2, 18, \mathbf{21}, 29, 36,$$

and the sample median is $M = 21$. If the number of observations is even, then the sample median is taken to be the average of the two middle values. For example, if we observe the values

$$3, 12, \mathbf{18, 28}, 32, 59,$$

the sample median is

$$M = \frac{(18 + 28)}{2} = 23.$$

For the study about the desired number of sexual partners, the median is 1; this is in sharp contrast to the mean, which is 64.3.

Mathematicians have long realized that the sample mean can be unsatisfactory for a variety of reasons, one of which is that it has the lowest possible finite sample breakdown point. It might seem that when probability curves are exactly symmetric, the sample mean is satisfactory, but this is often not the case. For example, nearly two hundred years ago, Laplace realized that in some situations the sample median tends to be a more accurate estimate of the central point (the population mean μ), than the sample mean, $\bar{X}$. This is *not* a compelling argument for abandoning the sample mean in favor of the sample median, however, because in other situations the mean is more accurate. But for now we continue to focus on the notion of the breakdown point because it has practical value for some of the applied problems we will address.

Notice that the sample median involves a type of trimming. For five observations we eliminate the two smallest and two largest values to get the median. For six values, we again trim the two largest and smallest, and then we average the values that remain.

Now consider the finite sample breakdown point of the median. How many outliers does it take to make the sample median arbitrarily large? Looking at our two numerical examples, we see that if we increase the two largest of the five values, this has no influence on the value of the sample median—we need a minimum of three outliers to make the sample median arbitrarily large. In a similar manner, decreasing the two smallest values does not affect the sample median either. In general, if we have n observations, the minimum number of outliers needed to make the sample median arbitrarily large is approximately $n/2$. In a similar manner, the number needed to make the sample median arbitrarily small is again $n/2$, so the finite sample breakdown point is approximately 1/2. The median achieves the highest possible breakdown point as opposed to the sample mean, which has the lowest breakdown point. For some purposes the breakdown point turns out to be of paramount importance, but for other goals, the criterion of having a high breakdown point must be tempered by other considerations to be covered.

A WEIGHTED MEAN

There is a generalization of the sample mean that turns out to be important for some purposes. It is called a *weighted mean* and it simply refers to multiplying (or *weighting*) each of the observations by some constant and adding the results. When the goal is to estimate the population mean, the weights are typically chosen so that they sum to one, but as we will see, there are other common goals where the weights do not sum to one.

Consider again the five values

$$2, 18, 21, 29, 36.$$

If we multiply each value by 1/5 and sum the results, we get

$$\frac{1}{5}(2) + \frac{1}{5}(18) + \frac{1}{5}(21) + \frac{1}{5}(29) + \frac{1}{5}(36) = 21.2,$$

which is just the usual sample mean. In this particular case, each weight has the same value, namely 1/5.

Here is another example of a weighted mean:

$$.3(2) + .1(18) + .2(21) + .2(29) + .2(36) = 19.6.$$

The weights in this case are .3, .1, .2, .2, and .2, and as is evident, they sum to one. This particular weighted mean provides an estimate of the population mean that differs from the sample mean, 21.2. There are, of course, infinitely many choices for the weights that sum to one. At some level it might seem obvious that among all the weighted means we might consider, the optimal estimate of the population mean is obtained with the sample mean where each observation receives the same weight. There are precise mathematical results describing circumstances under which this speculation is correct. However, for many other applied problems, it is not immediately obvious which choice of weights is optimal, but our only goal here is to simply introduce the notion of a weighted mean.

Provided that all the weights differ from zero, the finite sample breakdown point of the weighted mean is the same as it was for the sample mean, $1/n$. That is, a single unusual value can make the weighted mean arbitrarily large or small.

VARIANCE

A general problem of extreme importance in applied work is measuring the dispersion among a batch of numbers. Many more than one hundred such measures have been proposed, but one particular method turns out to be especially convenient when working with the mean. It is called the *population variance* and is typically labeled σ^2.

Imagine you are interested in the typical anxiety level of all adults and that anxiety is measured on a ten-point scale, meaning that possible anxiety scores range over the integers 1 through 10. Further imagine that if we could measure all adults, the average score would be $\mu = 6$. The population variance of the anxiety scores, σ^2, is the average squared distance between μ and the anxiety

VARIANCE

scores of all adults if they could be measured. Mathematicians write this more succinctly as

$$\sigma^2 = E[(X - \mu)^2], \tag{2.1}$$

where E means expected value. Here, X represents the anxiety score of an adult. There is a more formal definition of expected value, but the details are relegated to Appendix A. Readers not interested in these details can read E as the average we would get if all individuals of interest could be measured. For example, $E(X)$ is read as the expected value of X and is just the average value of all anxiety scores, which we call the population mean. That is, $\mu = E(X)$. In a similar manner, among all individuals we might measure, the right side of Equation 2.1 refers to the expected or average value of the squared difference between an individual's anxiety score and the population mean.

In practice the population variance is rarely (if ever) known, but it can be estimated from the sample of observations available to us. Imagine we have ten randomly sampled adults with anxiety scores

$$1, 4, 3, 7, 8, 9, 4, 6, 7, 8.$$

We do not know the population mean, but we can estimate it with the sample mean, which in the present situation is $\bar{X} = 5.7$. Then an estimate of the population variance is obtained by computing the average squared difference between 5.7 and the observations available to us. That is, use

$$\frac{1}{10}\{(1 - 5.7)^2 + (4 - 5.7)^2 + \cdots + (8 - 5.7)^2\}.$$

However, following a suggestion by Gauss, a slightly different estimate is routinely used today. Rather than divide by the number of observations, divide by the number of observations minus one instead. In a more general notation, if we observe the n values $X_1, X_2, \ldots, X_n$, use

$$s^2 = \frac{1}{n-1}\{(X_1 - \bar{X})^2 + \cdots + (X_n - \bar{X})^2\}. \tag{2.2}$$

The quantity s^2 is called the *sample variance* to distinguish it from the population variance. In the illustration, $s^2 = 6.68$, and this is used as an estimate of the population variance σ^2. The square root of the population variance, σ, is called the population *standard deviation* and is estimated with s, the square root of s^2. So in the illustration the estimate of the population standard deviation is $s = \sqrt{6.68} = 2.58$.

Now consider the finite sample breakdown point of the sample variance. In the current context, this means we want to know the minimum proportion of outliers required to make the sample variance arbitrarily large. The answer is $1/n$, the same result we got for the sample mean. That is, a single outlier can completely dominate the value of the sample variance. For example, if we observed the values 5, 5, 5, 5, 5, 5, 5, 5, 5, and 5, then $s^2 = 0$; there is no variation among these ten values. But if the last value were 10 instead, $s^2 = 2.5$, and if it were 100, $s^2 = 902.5$. The low breakdown point of the mean is a practical concern in some applied settings, but the low breakdown point of the variance turns out to be especially devastating—even when observations are symmetrically distributed around some central value.

MEASURING ERROR

Our next goal is to relate the mean and median to a particular notion of error that plays a fundamental role in basic methods. This will help set the stage for a general approach to characterizing the typical individual under study, as well as finding optimal methods for studying the association among two or more variables.

To be concrete, consider a game where two contestants are asked to guess the height of the next five people they will meet. Each contestant is allowed one guess that will be applied to each of these five individuals. Imagine that the first contestant guesses 68 inches and the second guesses 69. Further imagine that the heights of the first five people turn out to be 64, 65, 67, 74, and 80 inches. Which contestant won?

The answer depends on how we measure the accuracy of each guess. First consider the squared difference between the guess made and the values observed. Let's start with the first contestant, who guessed 68 inches. For the person who happened to be 64 inches tall, the accuracy of the guess is measured by $(64 - 68)^2 = 16$, the squared difference between the two values. For the person who was 74 inches tall, the error is $(74 - 68)^2 = 36$. Now suppose we compute the squared error for each of the five individuals and sum the results to get an overall measure of how well the first contestant performed. We find that the sum of the squared errors is 206. Next we repeat this process for the second contestant who guessed 69 inches. For the individual who was 64 inches, the error is now given by $(64 - 69)^2 = 25$. In this particular case, the first contestant did better because her error was only 16. However, computing the squared error for each of the five weights and adding the results, we get 191 for the second contestant. This is less than the sum of the squared errors for the first contestant, so we declare the second contestant the winner.

But why did we use squared error? Would it make a difference if we used absolute error instead? Let's try that. For the first contestant, who guessed 68 inches, let's compute the absolute value of the error made for the person who was 64 inches tall. Now we get $|64 - 68| = 4$. If we do this for the other four measures and we add the results, we get 26, which measures the overall accuracy based on the guess made by the first contestant. Repeating this for the second contestant, we get 27. Now the first contestant is the winner, not the second! The overall accuracy of the first one's guesses is better when we use absolute values!

Let's ask another question. For the five heights we got, what would have been the best guess, the value that would minimize the sum of the squared errors? The answer is the mean, which is 70. If one of the contestants had guessed 70 inches, they could not be beaten by the other contestant. Choosing as our guess the value that minimizes the sum of the squared errors is an example of what is called the *least squares principle*. For the problem at hand, where the goal is to minimize the error when guessing the height of five individuals, the least squares principle leads to the mean.

What if we use absolute values instead? If we sum the absolute values of the errors to get an overall measure of accuracy, what would have been the best guess? The answer is the median, or middle value, which is 67.

It helps to consider a slight modification of our guessing game. Now imagine that we are given the values 64, 65, 67, 74, and 80, and the goal is to choose a single number close to these five values. Let's temporarily call this number c. If by *close* we mean the sum of the squared distances, then the closeness of c to the five values at hand is

$$(64 - c)^2 + (65 - c)^2 + (67 - c)^2 + (74 - c)^2 + (80 - c)^2.$$

To minimize this last expression, viewed as a function of c, it can be seen that c must satisfy

$$(64 - c) + (65 - c) + (67 - c) + (74 - c) + (80 - c) = 0.$$

A little algebra shows that c is just the mean of the five numbers. More generally, for any batch of numbers, the sample mean ($\bar{X}$) minimizes the sum of the squared distances. If, however, we use

$$|64 - c| + |65 - c| + |67 - c| + |74 - c| + |80 - c|$$

to measure closeness, this leads to taking c to be the median.

In the illustration, the height of five individuals was used, but there is nothing special about the number five. If we had a hundred or a thousand individuals, a similar result would apply. That is, under squared error, the optimal guess would be the mean, but if we use absolute error, the median should be used instead.

There are infinitely many ways we can measure closeness in addition to the two methods used here. For example, we could use the absolute differences raised to the power a, where a is any number greater than zero. Squared error is being used when $a = 2$ and absolute value corresponds to $a = 1$. In 1844, Ellis suggested an even broader class of methods for measuring closeness, which covers what we now call M-estimators of location. The sample mean is included as a special case. In our current context it is completely arbitrary which measure is used, an issue that concerned both Laplace and Gauss, so other criteria must be invoked to achieve some type of resolution. But before turning to this important topic, we extend the notion of error to the problem of fitting a line to a scatterplot of points.

FITTING A STRAIGHT LINE TO DATA

Next we consider the extension of the least squares principle to the problem of fitting a line to a plot of points. To provide some historical sense of why least squares came to dominate in applied work, we describe a classic problem of the eighteenth century.

Newton, in his *Principia*, argued that the rotation of the earth should cause it to be flattened somewhat at the poles, and it should bulge at the equator. (That is, rotation should produce an oblate spheroid.) In contrast, Domenico Cassini, director of the Royal Observatory in Paris, took the opposite view: The earth is flattened at the equator. (It is a prolate spheroid.) Two methods were employed in an attempt to determine which view is correct: pendulum experiments and arc measurements. Pendulum experiments by Richer in 1672 had already suggested that the earth is not a perfect sphere. This speculation was based on observations that pendulums at the equator are less affected by gravity than pendulums in Paris. Both types of experiments presented similar mathematical difficulties in terms of resolving the oblate versus prolate speculations, but here we focus on arc measurements.

The idea was to measure the linear length of a degree of latitude at two or more different places. It was reasoned that if a degree near the equator was found to be shorter than one near the pole, then the shape of the earth is oblate, and the difference between two measurements could be used to measure the oblateness. Measurements by Cassini and his son Jacques, made before 1720,

Fitting a Straight Line to Data

supported the prolate-spheroid view, but concerns about the accuracy of the measurements precluded a definitive conclusion about which view is correct.

A portion of the analysis hinged on being able to establish the relationship between arc length and latitude. When using short arcs, a simple linear relationship can be assumed to exist between a certain transformation of the latitude (X) and arc length (Y). That is, the relationship between X and Y has the form

$$Y = \beta_1 X + \beta_0, \tag{2.3}$$

and the problem is to determine the values for the unknown slope β_1 and the unknown intercept β_0 based on observations made. Now, any two distinct points determine a line. In the current context, this means that if we are given two pairs of observations for latitude and arc length, we can determine the slope and the intercept. For example, if for the transformed latitude $X = 0$ we find that the arc length is $Y = 56,751$, and at the transformed latitude $X = 0.8386$ we find that $Y = 57,422$, then the slope is given by

$$\frac{57,422 - 56,751}{0.8386 - 0} = 800.1.$$

Once this is done, the issue of the shape of the earth can be resolved by computing $\beta_1/3\beta_0$, which measures its ellipticity. According to Newton, this ratio should be about 1/230.

Over a period of fifteen years, Roger Boscovich attempted to resolve the shape of the earth using the data in Table 2.1. Boscovich had two problems. First, any two points can be used to determine the slope and intercept, but he had five points, and he got different results depending on which two of the five points he selected. In effect, the five points yielded ten estimates of the slope and intercept, all of which differed from one another. Second, the discrepancies were due, at least in part, to measurement errors. So the issue was whether the data could be combined in some manner to resolve the shape of the earth. A basic concern is whether the effects of measurement errors are exacerbated or reduced when combining the data.

One possibility is to compute the slope for all pairs of points and average them, an idea that dates back to at least 1750. Here there are ten pairs of points one could use. The ten corresponding slopes, written in ascending order, are

$$-349.19, 133.33, 490.53, 560.57, 713.09, 800.14, 852.79, 957.48, 1185.13, 1326.22.$$

The average of these ten slopes is 667. In contrast, the median is 757.6, so an obvious issue is which of these two estimates can be expected to be more

TABLE 2.1 • BOSCOVICH'S DATA ON MERIDIAN ARCS

PLACE	TRANSFORMED LATITUDE	ARC LENGTH
Quito	0.0000	56,751
Cape of Good Hope	0.2987	57,037
Rome	0.4648	56,979
Paris	0.5762	57,074
Lapland	0.8386	57,422

accurate, a problem we will turn to later. Yet another practical problem, particularly during the precomputer age, was that the notion of computing all possible slopes becomes impractical as the number of observations, or the number of predictors, increases. For example, if Boscovich had fifty observations, the number of slopes to be computed and averaged would be 1225, and for a hundred pairs of points it would be 4950.

What other numerical method might be used to choose one particular line over the infinitely many lines one might use? Boscovich got an idea for how to approach this problem that Laplace would later declare to be ingenious. His idea was to use the error or discrepancy between a proposed line and the data, a suggestion that represents a slight generalization of how we measured error earlier in this chapter.

Figure 2.5 illustrates this approach using the data in Table 2.1. The line in Figure 2.5 connects the two most extreme points and represents Boscovich's first attempt at fitting a line to the data. The equation for this line is

$$\hat{Y} = 800.14X + 56,751,$$

where the notation $\hat{Y}$ is used to make a distinction between the predicted arc length ($\hat{Y}$) versus the arc length we observe (Y). Consider the point indicated in Figure 2.5, where the transformed latitude is 0.5762. From the equation used by Boscovich, the predicted arc length ($\hat{Y}$) is 57,212, but the observed arc length is 57,074, so we have a discrepancy of $57,074 - 57,212 = -138$.

The discrepancy just illustrated is an example of what we now call a *residual*. Notice that for every observed arc length (Y), we have a corresponding predicted arc length ($\hat{Y}$) and a resulting discrepancy. In a more formal notation, a residual is

$$r = Y - \hat{Y}.$$

In our example, there are five residuals, one for each point (or pair of observations) available to us. Boscovich's idea is, when fitting a line to data, choose

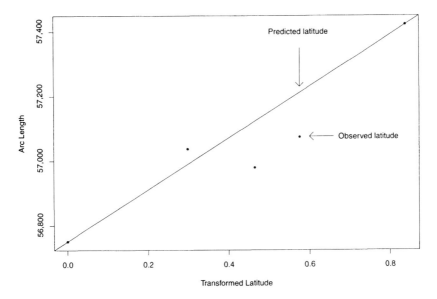

FIGURE 2.5 • A graphical illustration of a residual using Boscovich's data in Table 2.1. The straight line represents Boscovich's first attempt at fitting a line to the data. For the transformed latitude $X = .5762$, the observed arc length is $Y = 57,074$. The predicted arc length according to the straight line considered by Boscovich is $\hat{Y} = 800.1(.5762) + 56,751 = 57,212$. The corresponding residual is $57,074 - 57,212 = -138$.

the line that minimizes the sum of the absolute residuals, and he devised a numerical scheme for implementing the method. For the special case where the slope is zero, Boscovich's method leads to using the median to estimate the intercept.

In 1809, Legendre published a paper suggesting that one minimize the sum of the squared residuals instead. That is, use the least squares principle already discussed. When the slope is zero, this leads to using the sample mean to estimate the intercept. This simple modification of Boscovich's approach would prove to have great practical value, but as we shall see, it also lays the foundation for disaster in many applied settings. For now we focus on the positive aspects of the method. The first is that in the hands of Gauss, a computationally feasible method for determining the slope and intercept can be derived. In fact, Gauss devised his so-called method of elimination that made it possible to handle multiple predictors. Second, it provided a framework for mathematically treating some problems of considerable practical importance, which we will cover in later chapters. In fact, so valuable was Legendre's suggestion that it sparked the beginning of one of the greatest controversies in statistics: Who was the first to suggest using least squares? Gauss claimed to have been using it since 1795, but he was unable to provide compelling proof, and the issue remains unresolved to this day.

It should be noted that the problem considered by Boscovich is an example of what we now call simple linear regression, meaning that we have one predictor, usually labeled X (which represents the transformed latitude here), and an outcome variable Y (arc length). *Multiple regression* refers to a situation where there are two or more predictors.

TWO VIEWS OF THE COMPUTATIONS

There are two ways of viewing the computational aspects of the least squares regression method that will prove to be useful later. The first is that the estimated slope turns out be a weighted mean of the Y values. The weights are determined by the difference between the corresponding X values and the mean of all X values, as well as the variation among the X values.

To be more precise, consider again the data in Table 2.1, but with some additional information as indicated in Table 2.2. The sample mean of the transformed latitudes (the average of the X values) is $\bar{X} = 0.43566$. Subtracting this value from each of the five X values yields the results in the column headed by $X - \bar{X}$. If we square each value in this same column and add the results, we get 0.3915. Dividing each value in this column by 0.3915 yields the values in the column headed by W. More succinctly, and said in a slightly different manner,

$$W = \frac{X - \bar{X}}{(n-1)s_x^2},$$

where s_x^2 is the sample variance of the X values. The final column shows the values of the arc length Y multiplied by W. The sum of the values in this last column, 722, is a weighted mean of the Y values, the weights being given by the column headed by W. (Unlike the weighted mean previously discussed, the weights used to estimate the slope sum to zero, not one.) This weighted mean is the least squares estimate of the slope. (If more decimal places are retained, as would be done on a computer, the $W \times Y$ values would be altered slightly and the estimated slope would now be 723.4.)

As previously noted, weighted means have a finite sample breakdown point of $1/n$, meaning that a single outlier can have a large influence on their

FITTING A STRAIGHT LINE TO DATA

TABLE 2.2 • BOSCOVICH'S DATA ON MERIDIAN ARCS

Place	X	Y	$X - \bar{X}$	W	$W \times Y$
1	0.0000	56,751	−0.4357	−1.1127	−63,149
2	0.2987	57,037	−0.1370	−0.3498	−19,953
3	0.4648	56,979	0.0291	0.0744	4,240
4	0.5762	57,074	0.1405	0.3590	20,487
5	0.8386	57,422	0.4029	1.0292	59,097

value. Because the least squares estimate of the slope is just a weighted mean of the Y values, we see that it too has a finite sample breakdown point of only $1/n$. For example, a single unusual point, properly placed, can cause the least squares estimate of the slope to be arbitrarily large or small. This turns out to be of great practical importance, as we shall see in Chapters 9 and 10.

Here is another view of the computations that will be helpful. Recall that for each pair of points, we can compute a slope. For the data at hand, there are ten pairs of points yielding ten slopes, as previously indicated. For the first two pairs of points in Table 2.2—(0.0000, 56,751) and (0.2987, 57,037)—the slope is given by

$$\frac{57,037 - 56,751}{0.2987 - 0.0000} = 957.48.$$

Let $w = (0.2987 - 0.0000)^2 = 0.08922$ be the squared difference between the two X values (the transformed latitudes). In effect, least squares multiplies (weights) the original estimate of the slope, 957.48, by w. This process is repeated for all ten pairs of slopes, each slope being multiplied by the resulting value for w. Sum the results and call it A. Next, sum all of the w values and call it B. The least squares estimate of the slope is A/B; it is just a weighted mean of all the slopes corresponding to all pairs of points.

It is remarked that the computational method just described is never used in applied work when adopting the least squares principle—a method that requires substantially fewer computations is routinely used. It is the conceptual view that is important. The main point for now is that there are infinitely many weighted means one might use to estimate the slope (which in turn yields an estimate of the intercept). A fundamental issue is finding the weighted mean that is, on average, the most accurate. That is, the goal is to estimate the slope and intercept we would get if there were no measurement errors or if infinitely many observations could be made. What are the conditions under which the least squares method tends to be the most accurate?

Are there situations where it can be extremely inaccurate versus other weighted means we might use? The answer to the last question is an emphatic yes, as we shall see.

A SUMMARY OF KEY POINTS

- The sample mean, plus all weighted means (with all weights differing from zero), has the lowest possible finite sample breakdown point. That is, a single outlier can cause the mean and weighted mean to give a highly distorted indication of what the typical subject or measurement is like.

- The median has the highest possible finite sample breakdown, roughly meaning that it is relatively insensitive to outliers no matter how extreme they might be. But this is not a compelling argument for completely abandoning the mean for the median because of criteria considered in subsequent chapters.

- Two methods for measuring error were introduced, one based on squared error and the other based on absolute error. We get different ways of summarizing data depending on which measure we use. Squared error (the least squares principle) leads to the mean, but absolute error leads to the median.

- Least squares regression was introduced and shown to be a weighted mean, so it too can be highly influenced by a single outlier.

- The sample variance, s^2, also has a finite sample breakdown point of $1/n$. (This result provides one explanation for why modern robust methods, covered in Part II of this book, have practical value.)

CHAPTER 3

THE NORMAL CURVE AND OUTLIER DETECTION

No doubt the reader is aware that the normal curve plays an integral role in applied research. Properties of this curve, which are routinely described in every introductory statistics course, make it extremely important and useful. However, in recent years, it has become clear that this curve can be a potential source for misleading—even erroneous—conclusions in our quest to understand data. This chapter summarizes some basic properties of the normal curve that play an integral role in conventional inferential methods. But this chapter also lays the groundwork for understanding how the normal curve can mislead. A specific example covered here is how the normal curve suggests a frequently employed method for detecting outliers that can be highly misleading in a variety of commonly occurring situations. This chapter also describes the central limit theorem, which is frequently invoked in an attempt to deal with nonnormal probability curves. Often the central limit theorem is taken to imply that with about twenty-five observations, practical problems due to nonnormality become negligible. There are several reasons why this is erroneous, one of which is given here. The illustrations in this chapter provide a glimpse of additional problems to be covered.

THE NORMAL CURVE

The family of equations for the normal curve is

$$\frac{1}{\sqrt{2\pi}\sigma}e^{-(x-\mu)^2/(2\sigma^2)}, \quad (3.1)$$

where μ is the population mean around which observations are centered and σ is the population standard deviation, which is a measure of scale introduced in Chapter 2; it determines how tightly the curve is centered around the mean. (In Equation 3.1, e is Euler's constant and is approximately equal to 2.718.) All normal curves are bell-shaped and symmetric about the population mean. The normal curve with $\mu = 0$ and $\sigma = 1$ is called a *standard normal distribution*.

Figure 3.1 illustrates the effects of varying the mean (μ) and standard deviation (σ). The two normal curves on the left have the same mean (both are centered around zero), but they have different standard deviations. Notice that increasing the standard deviation from 1 to 1.5 results in a clear and noticeable change in the graph of the normal curve. (This property forms the basis of a common misconception discussed in Chapter 7.) The curve with the smaller standard deviation is more tightly centered around the population mean. If two normal curves have the same standard deviation but unequal means, the shapes of the curves are exactly the same; the only difference is that they are centered around different values.

There is a feature of the normal curve and the standard deviation that plays an integral role in basic inferential methods in statistics. For any normal curve with an arbitrary mean and standard deviation, the probability that an observation is within one standard deviation of the mean is exactly .68. For example, if a normal probability curve is centered around 100 (the population mean is $\mu = 100$) and its standard deviation is 4 ($\sigma = 4$), then the probability that a randomly sampled observation is between $100 - 4 = 96$ and $100 + 4 = 104$ is exactly .68. If the standard deviation is 8, then the probability of an observation being between $100 - 8 = 92$ and $100 + 8 = 108$ is again .68. This result generalizes to any mean. If the mean is 50 and the standard deviation is 10, then the probability of getting an observation between $50 - 10 = 40$ and $50 + 10 = 60$ is .68.

In a similar manner, the probability of being within two standard deviations of the population mean is exactly .954. For example, if again the mean is 100 and the standard deviation is 4, the probability that a randomly sampled observation is between $100 - 2(4) = 92$ and $100 + 2(4) = 108$ is exactly .954. In fact, for any multiple of the standard deviation, the probabil-

DETECTING OUTLIERS

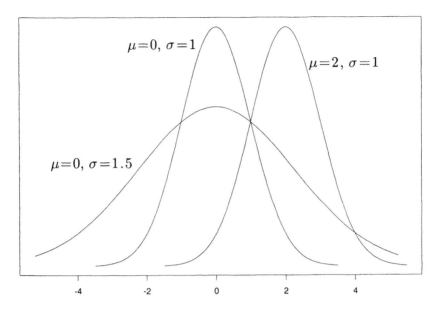

FIGURE 3.1 • These normal probability curves provide some sense of how such a curve changes when the mean and standard deviation are altered. Two of the normal distributions have equal means but unequal standard deviations. Note that increasing the standard deviation from 1 to 1.5 results in a clear and distinct change in a normal curve. Curves with equal standard deviations but unequal means are exactly the same, except that one is shifted to the right.

ity remains fixed. For example, the probability of being within 1.96 standard deviations of the mean is exactly .95. Figure 3.2 graphically illustrates this property.

A related result is that probabilities are determined exactly by the mean and standard deviation when observations follow a normal curve. For example, the probability that an observation is less than 20 can be determined exactly if we are told that the population mean is 22 and the standard deviation is 4. (The computational details are covered in virtually all introductory books on applied statistics.)

DETECTING OUTLIERS

The property of the normal curve illustrated in Figure 3.2 suggests a probabilistic approach to detecting outliers that is frequently employed: Declare a

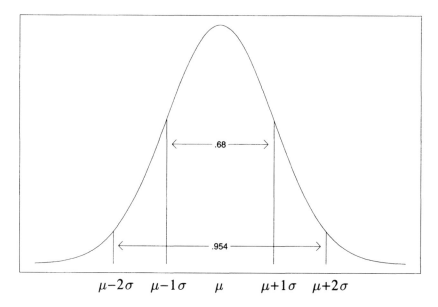

FIGURE 3.2 • For normal probability curves and any positive constant c, the probability that the distance of an observation from the population mean is less than $c\sigma$ is completely determined by c. That is, regardless of what the values for the population mean and variance might be, the constant c determines this probability. For example, the probability that an observation is within one standard deviation of the mean is exactly .68. That is, with probability .68, a randomly sampled observation will lie between $\mu - 1\sigma$ and $\mu + 1\sigma$. The probability that an observation is within two standard deviations of the mean is exactly .954.

value to be an outlier if it is more than two standard deviations from the mean. In symbols, declare the value X to be an outlier if

$$|X - \mu| > 2\sigma, \tag{3.2}$$

the idea being that there is a low probability that the distance of an observation from the mean will be greater than two standard deviations. For a normal probability curve, this probability is .046. In practice, the population mean and standard deviation are generally unknown, but they can be estimated as previously indicated, in which case the rule is to declare the value X an outlier if its distance from the sample mean is more than two sample standard deviations. That is, declare X an outlier if

$$|X - \bar{X}| > 2s. \tag{3.3}$$

Detecting Outliers

Unfortunately, this simple method for detecting outliers has a serious problem related to the finite sample breakdown point of $\bar{X}$ and s. To illustrate the problem, consider the values

$$2, 3, 4, 5, 6, 7, 8, 9, 10, 50.$$

The sample mean is 10.4 and the standard deviation is 14.15, so we see that the value 50 is declared an outlier because $|50 - 10.4|$ exceeds 2×14.15. But suppose we add another outlier by changing the value 10 to 50. Then $|\bar{X} - 50| = 1.88s$, so 50 would not be declared an outlier, yet surely it is unusual compared to the other values. If the two largest values in this last example are increased from 50 to 100, then $|\bar{X} - 100| = 1.89s$, and the value 100 still would not be declared an outlier. If the two largest values are increased to 1000, even 1000 would not be flagged as an outlier! This illustrates the general problem known as *masking*. The problem is that *both* the sample mean and the standard deviation are being inflated by the outliers, which in turn masks their presence.

In the illustrations, it might be thought that if we knew the population standard deviation, rather than having to estimate it with s, the problem of masking would no longer be relevant. It turns out that this speculation is incorrect. In fact, even when a probability curve appears to be normal, meaning that it is bell-shaped and symmetric about its mean, practical problems arise. (Details can be found in Chapter 7.)

BETTER METHODS

How can we get an effective method for detecting outliers that is not subject to masking? A key component is finding measures of location and scale that are not themselves affected by outliers. We know how to get a high breakdown point when estimating location: Replace the mean with the median. But what should be used instead of the sample standard deviation?

One choice that turns out to be relatively effective is the so-called *median absolute deviation statistic*, commonly referred to as MAD. It is computed by subtracting the median from each observation and then taking absolute values. In symbols, compute

$$|X_1 - M|, |X_2 - M|, \ldots, |X_n - M|.$$

The median of the n values just computed is MAD.

Here is an illustration using the values 2, 4, 6, 8, 9, 12, and 16. The median is $M = 8$. Subtracting 8 from each of the seven values and then taking absolute values, yields

$|2-7| = 5, |4-7| = 3, 1, 0, 1, 4, 8.$

The median of the seven values just computed is MAD; that is, MAD= 3.

For normal probability curves, it turns out that MAD/.6745 estimates the population standard deviation, σ. Simultaneously, the sample median, M, estimates the population mean, μ. For many purposes, using MAD to estimate the population standard deviation is unsatisfactory. For one, it tends to be a less accurate estimate than the sample standard deviation s when observations do follow a normal curve. However, MAD is much less sensitive to outliers; its finite sample breakdown point is approximately .5, the highest possible value, so it is well suited for detecting outliers.

We can modify our rule for detecting outliers in a simple manner: Declare the value X an outlier if

$$|X - M| > 2\frac{\text{MAD}}{.6745}. \tag{3.4}$$

As an illustration, consider again the study dealing with the desired number of sexual partners by young males, but to make the illustration more salient, we omit the value 6000. A portion of the data, written in ascending order, looks like this:

$$0, 0, 0, 0, 0, 1, \ldots, 30, 30, 40, 45, 150, 150.$$

The sample mean is 7.79, and the standard deviation is 21.36. If we use our outlier detection rule based on the mean and standard deviation, we see that the value 150 is declared an outlier, but the other values are not. In contrast, using our rule based on the median and MAD, all values greater than or equal to 4 are declared outliers. That is, forty-one values are declared outliers versus only the value 150 when using the mean and standard deviation.

THE BOXPLOT

Another method for detecting outliers is briefly mentioned because it is frequently employed and recommended. It is based on a graphical method for summarizing data called a *boxplot*, an example of which is shown in Figure 3.3. Construction of a boxplot begins by computing what are called the lower and upper quartiles, but the computational details are not covered here. (There are, in fact, at least a half dozen methods for computing quartiles.) The important point is that the quartiles are defined so that the middle half of the observed values lie between them. In Figure 3.3, the lower and upper quartiles are approximately 7 and 15, respectively, so about half of the values used to construct the boxplot lie between 7 and 15. (In addition, the lower fourth of the values are less than 7, and the upper fourth are greater than 15.) The boxplot uses the difference between the upper and lower quartiles, called the *interquartile range*, as a measure of dispersion, which in turn plays a role in deciding whether an observation is an outlier. If a value exceeds the upper quartile plus 1.5 times the interquartile range, it is declared an outlier. (In symbols, if F_U and F_L are the upper and lower quartiles, $F_U - F_L$ is the interquartile range, and a value is declared an outlier if it is greater than $F_U + 1.5(F_U - F_L)$.) Similarly, a value is declared an outlier if it is less than the lower quartile minus 1.5 times the interquartile range. (That is, a value is an outlier if it is less than $F_L - 1.5(F_U - F_L)$.) The lines extending out from the box in Figure 3.3 are called *whiskers*. The ends of the whiskers mark the smallest and largest values not declared outliers. So points lying beyond the end of the whiskers are declared outliers. In Figure 3.3, no outliers are found. Figure 3.4 shows a boxplot where two values are declared outliers.

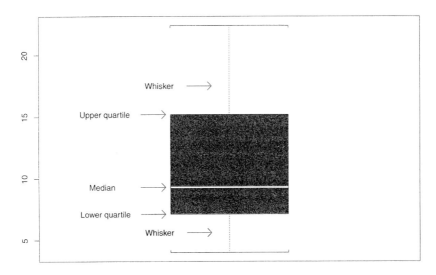

FIGURE 3.3 • An example of a boxplot. Construction of a boxplot begins by computing the lower and upper quartiles, which are defined so that approximately the middle half of the values lie between them. So, about 25 percent of the values plotted are less than the lower quartile, which in this case is approximately 7. Similarly, about 25 percent of the values are greater than the upper quartile, which is approximately 15.

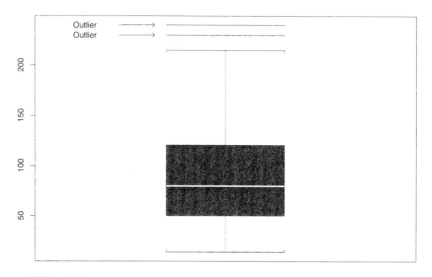

FIGURE 3.4 • Another boxplot, but in contrast to Figure 3.3, two values are flagged as outliers.

The boxplot has a finite sample breakdown point of .25, meaning that more than 25 percent of the values must be outliers before the problem of masking arises. For most situations, it seems that a finite sample breakdown point of .25 suffices, but exceptions might occur. For the data on the desired number of sexual partners, using the median and MAD led to declaring 41 values as outliers, and this is about 39 percent of the hundred five values. If a boxplot is used, values greater than or equal to 15—about 10 percent of the values—are declared outliers.

Research on outlier detection methods continues, a few additional issues will be discussed later, but a more detailed discussion of outlier detection goes beyond the main thrust of this book.

THE CENTRAL LIMIT THEOREM

There is a rather remarkable connection between the sample mean and the normal curve based on the central limit theorem derived by Laplace. Let's suppose we are interested in feelings of optimism, among all adults living in France. Imagine we have some measure of optimism, and as usual let μ

and σ^2 represent the population mean and variance. Further imagine that we randomly sample twenty adults and get a sample mean of 22, so our estimate of the population mean is 22. But suppose a different team of researchers randomly samples twenty adults. Of course, they might get a different sample mean from what we got; they might get 26. If yet another team of researchers sampled twenty adults, they might get yet another value for the sample mean. If this process could be repeated billions of times (and hypothetically infinitely many times), each time yielding a sample mean based on twenty observations, and if we plotted the means, what can be said about the plotted means?

Laplace found that provided each mean is based on a reasonably large sample size, the plots of the means will approximately follow a normal curve. In fact, the larger the number of observations used to compute each sample mean, the better the approximation. Moreover, this normal curve would be centered around the population mean. That is, the sample means would tend to be centered around the value they are trying to estimate. Furthermore, the variance of the normal curve that approximates the plot of the sample means is determined by the population variance (σ^2) and the number of observations used to compute the mean. If observations follow a probability curve having population variance six, and if again the sample means are based on twenty observations, the variation of the sample means is exactly 6/20. More generally, if n observations are randomly sampled from a curve having population variance σ^2, the variance of the normal curve that approximates the plot of sample means will be σ^2/n.

The phrase *reasonably large* is rather vague. How many observations must be used when computing the sample means so that there is fairly good agreement between the plot of the means and a normal curve? There is no theorem giving a precise answer; we must rely on experience, although theory suggests where to look for problems.

We begin with a standard illustration of the central limit theorem, where observations follow a so-called uniform distribution. That is, the probability curve is as shown in Figure 3.5. This curve says that all values lie somewhere between 0 and 1, and all values are equally likely. As is evident, the curve in Figure 3.5 looks nothing like a normal curve. The population mean for this curve is .5, and the variance is 1/12, so the central limit theorem says that if each mean is based on n values and n is sufficiently large, a plot of the means will be approximately normal and centered around .5 with variance $1/12n$.

Now imagine we randomly sample twenty values and compute the mean. We might get .68. If we sample a new set of twenty values, this time we might get .42. Of course, we cannot repeat this process infinitely many times, but

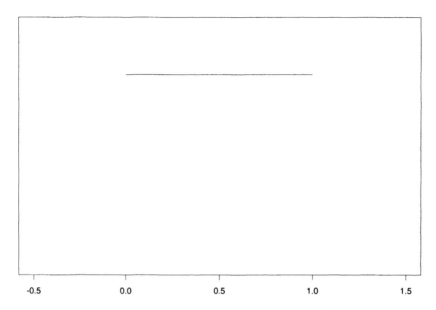

FIGURE 3.5 • A graphical depiction of the so-called uniform distribution. This probability curve is often used to illustrate the central limit theorem. Note that it looks nothing like a normal curve, yet plots of means are approximately normal when a sample size of only twenty is used to compute each mean.

we can get a fairly accurate sense of what the plot of infinitely many sample means would look like by repeating this process four thousand times with a computer and plotting the results. Figure 3.6 shows a plot of the four thousand means plus the curve we would expect based on the central limit theorem. As can be seen, there is fairly good agreement between the normal curve and the plot of the means, so in this particular case the central limit theorem gives reasonably good results with only twenty observations used to compute each mean.

Let's try another example. We repeat our computer experiment, but this time we sample observations having the probability curve shown in Figure 3.7 (which is called an *exponential distribution*). Again this curve looks nothing like a normal curve. Both its mean and variance are 1, so the central limit theorem suggests that plots of means will be centered around 1 with variance $1/n$. Figure 3.8 shows the plot of four thousand means generated on a computer. Again, with only twenty values used to compute each mean, the normal curve provides a reasonably good approximation of the plotted sample means.

So we have two examples where we start with a probability curve that

The Central Limit Theorem

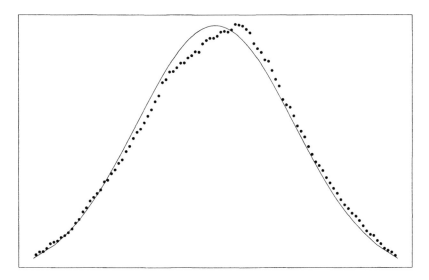

FIGURE 3.6 • A plot of four thousand means versus the predicted curve based on the central limit theorem when observations are sampled from a uniform distribution. In this case, a normal curve provides a good approximation of the plotted means with only twenty observations used to compute each mean.

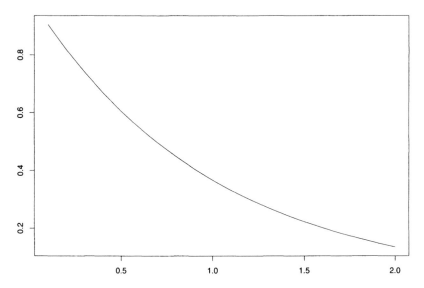

FIGURE 3.7 • A graphical depiction of the so-called exponential distribution. This is another probability curve often used to illustrate the central limit theorem.

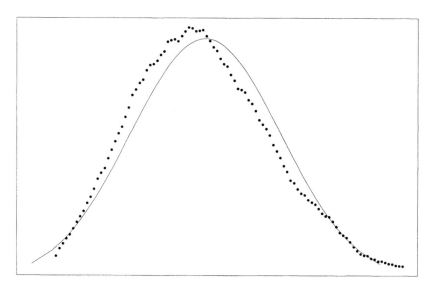

FIGURE 3.8 • A plot of four thousand means versus the predicted curve based on the central limit theorem when observations are sampled from an exponential distribution. Again the normal curve suggested by the central limit theorem provides a good approximation of the plotted means with only twenty observations used to compute each mean.

looks nothing like a normal curve, but plots of means are approximately normal when each mean is based on only twenty values. This might suggest that in general the central limit theorem applies with small sample sizes, but there are at least two problems. A description of one of these problems must be postponed until Chapter 5. To illustrate the other, let's consider an example where the probability curve is as shown in Figure 3.9. When twenty observations are used to compute each sample mean, the plot of the means is poorly approximated by a normal curve, particularly in the left tail, as indicated in Figure 3.10. If we increase the number of observations used to compute each mean, then according to the central limit theorem the approximation will improve. But if we use fifty observations to compute each sample mean, the approximation remains poor. If each mean is based on one hundred observations, the plot of means is now reasonably well approximated by a normal curve. So we see that in some cases, with twenty observations we get a good approximation, but there are situations where we need about a hundred observations instead.

The Central Limit Theorem

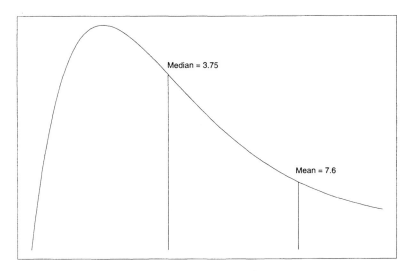

FIGURE 3.9 • An example of an asymmetric probability curve for which outliers are relatively common. Experience indicates that such curves are common in applied work.

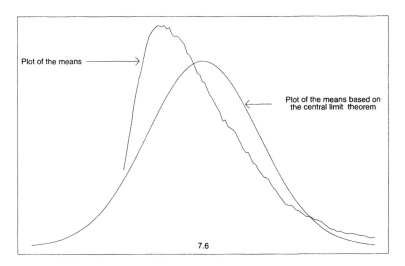

FIGURE 3.10 • A plot of four thousand means versus the predicted curve based on the central limit theorem when observations are sampled from the distribution shown in Figure 3.9. The normal curve for approximating the plot of the means, suggested by the central limit theorem, performs poorly when twenty observations are used to compute each mean.

In Figure 3.9, which is just a reproduction of Figure 2.4, the population mean is in a fairly extreme portion of the right tail, as noted in Chapter 2. Observe that despite this, the sample means are centered around this extreme value. That is, the sample means satisfy their intended goal of estimating the population mean, even though the population mean might be far removed from most of the observations.

What distinguishes the three illustrations of the central limit theorem? The first two are based on probability curves characterized by what are called *light tails*. This roughly means that outliers tend to be rare when sampling observations. In the last example, sampling is from a heavy-tailed probability curve where outliers are common—a situation that occurs frequently in applied work. So a tempting generalization is that we can assume sample means follow a normal curve if sampling is from a light-tailed probability curve, even when each mean is based on only twenty observations, but the approximation might be poor if sampling is from an asymmetric probability curve that is likely to produce outliers. However, there is one more problem that must be taken into consideration, but we must cover other details before it is described. For now, we will merely remark that even when sampling from a light-tailed probability curve, practical problems arise in some situations. (The details are explained in Chapter 5; see in particular the text regarding Figures 5.5 and 5.6.)

NORMALITY AND THE MEDIAN

Mathematical statisticians have derived many measures of location in addition to the mean and median. There is a version of the central limit theorem that applies to most of them, including all the measures of location considered in this book. For example, if we sample twenty values and compute the median, and if we repeat this process billions of times, the plot of the medians will be approximately normal, provided each median is based on a reasonably large number of observations. For symmetric probability curves, such as the normal, these medians will be centered around the population mean, which is equal to the population median. For asymmetric probability curves, the sample medians will be centered around the population median, which in general is not equal to the population mean, as explained in Chapter 2.

Let's repeat our computer experiment where we sampled values having the uniform probability curve shown in Figure 3.5, but now we compute me-

Normality and the Median

dians rather than means. A plot of the medians appears in Figure 3.11, and again we get a good approximation with a normal curve.

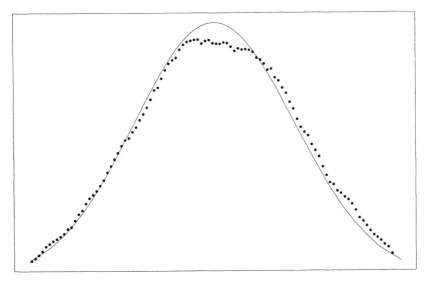

FIGURE 3.11 • There is a version of the central limit theorem that applies to medians and weighted means. This figure shows that when sampling observations from the uniform distribution in Figure 3.5, the plotted medians are well approximated by a normal curve when each median is based on only twenty observations.

Now look at Figure 3.12, which shows what we get when sampling from the probability curve in Figure 3.9. The central limit theorem says that the sample medians should be centered around the population median (3.75). We see that even when each median is based on twenty observations, the central limit theorem provides a much better approximation of the plot of medians than the plot of means. Put another way, convergence to normality is quicker when using medians. Roughly, practical problems with the central limit theorem, due to outliers, are more likely when the finite sample breakdown point is low. As previously indicated, the sample mean has the lowest possible finite sample breakdown point (only one outlier can completely dominate its value) compared to the median, which has the highest possible value, .5. So, in situations where inferences are based on the central limit theorem (using methods to be covered), larger sample sizes are needed to avoid practical problems with means than when using medians.

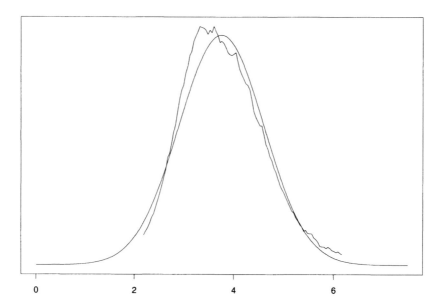

FIGURE 3.12 • Unlike means, when sampling observations from the asymmetric, heavy-tailed distribution in Figure 3.9, the normal curve used to approximate the plotted medians is in fairly close agreement with the four thousand medians generated on a computer.

Versions of the central limit theorem also apply when dealing with regression. Imagine repeating an experiment infinitely many times, each time randomly sampling n points and computing the slope. Chapter 2 pointed out that the least squares estimate of the slope of a regression line can be viewed as a weighted mean of the outcome (Y) values. This suggests that we would get a good approximation of the plotted slopes if in each experiment a reasonably large number of pairs of observations is used, and this speculation turns out to be correct under fairly general conditions. Again there is the issue of how many points need to be sampled to get good results with the central limit theorem in applied work. For now, suffice it to say that this issue turns out to be nontrivial.

Three points should be stressed before concluding this chapter. First, it is not suggested that medians should be routinely used instead of means. Although in some instances the median converges to normality more quickly than the mean, there are practical problems with medians, which will be described in Chapter 5. What is needed is some measure of location that performs about as well as the mean when the probability curve is normal but continues to perform well when outliers are common.

Second, a tempting strategy is to check for outliers and use means if none are found, but this can lead to serious practical problems for reasons that will be described in Chapter 5.

Third, if we were to repeat our computer experiment by sampling observations from a symmetric, heavy-tailed probability curve, it would appear that the central limit theorem performs well with means using only twenty observations to compute each mean. There are, however, two serious problems, as we shall see.

A SUMMARY OF KEY POINTS

- For all normal probability curves and any constant c, the probability that an observation does not differ from the population mean by more than $c\sigma$ is completely determined by c. For example, if $c = 1$, then with probability .68, $|X - \mu| < \sigma$. This property of normal curves suggests a commonly used rule for detecting outliers: Declare the value X an outlier if it satisfies Equation 3.3. This rule can be highly unsatisfactory, however, due to masking.

- An outlier detection rule that avoids the problem of masking is given by Equation 3.4. It avoids masking because it is based on measures of location and scale that have a high finite sample breakdown point. A boxplot also uses an outlier detection rule that reduces masking.

- The central limit theorem says that with a sufficiently large sample size, it can be assumed that the sample mean has a normal distribution. In some cases a good approximation of the sample mean is obtained with $n = 20$. But it was illustrated that in other situations, $n = 100$ is required. (In subsequent chapters we will see that even $n = 160$ may not be sufficiently large.)

- A version of the central limit theorem applies to the sample median. It was illustrated that situations arise where the distribution of the median approaches a normal curve more quickly, as the sample size increases, than does the distribution of the mean.

CHAPTER 4

ACCURACY AND INFERENCE

There is a classic result in statistics called the Gauss-Markov theorem. It describes situations under which the least squares estimator of the slope and intercept of a regression line is optimal. A special case of this theorem describes conditions under which the sample mean is optimal among the class of weighted means. In order to justify any competitor of least squares, we must understand the Gauss-Markov theorem and why it does not rule out competing methods such as the median, and other estimators to be described later. Therefore, one goal in this chapter is to give a relatively nontechnical description of this theorem. Another goal is to introduce the notion of a confidence interval, a fundamental tool used to make inferences about a population of individuals or things. We will see that a so-called homoscedastic error term plays a central role in both the Gauss-Markov theorem and the conventional confidence interval used in regression. Homoscedasticity turns out to be of crucial importance in applied work because it is typically assumed and because recent results indicate that violating the homoscedasticity assumption can be disastrous. Fortunately, effective methods for dealing with this problem have been devised.

SOME OPTIMAL PROPERTIES OF THE MEAN

We begin by temporarily stepping back in time and, as was typically done, assuming observations are symmetrically distributed about the population mean. For illustrative purposes, imagine we want to know the typical cholesterol level of individuals taking a new medication. When we estimate the population mean with a sample of individuals, we make an error—there will be some discrepancy between the sample mean and the population mean. If the population mean is $\mu = 230$, for example, typically the sample mean will differ from 230. Because it is assumed that sampling is from a symmetric probability curve, we could estimate the population mean with the sample median instead. Of course, it will also differ from the population mean in most cases. Which of these two estimators ($\bar{X}$ or M) is, in general, closer to the true population mean? Is there some other method of estimation that will give more accurate results on average? What are the general conditions under which the sample mean is more accurate than any other estimator we might choose?

Both Laplace and Gauss made major contributions toward resolving these questions. Laplace's approach was based on the central limit theorem. That is, he did not assume that values are sampled from a normal curve, but he did assume that sample sizes are large enough that the plot of the means, for example, would follow a normal curve to a high degree of accuracy. Gauss derived results under weaker conditions: He assumed random sampling only—his results do not require any appeal to the normal curve.

Our immediate goal is to describe how well the sample mean estimates the population mean in terms of what is called mean squared error. Henceforth, we will not assume observations are symmetrically distributed about the mean.

Imagine we conduct an experiment and get a sample mean of 63 and a second team of researchers conducts the same experiment and gets a mean of 57. Then imagine that infinitely many teams of researchers conduct the same experiment. Generally, there will be variation among the sample means. Of course, some sample means will be closer to the population mean than others. The *mean squared error* of the sample mean refers to the average or expected squared difference between the infinitely many sample means and the population mean. In symbols, the mean squared error of the sample mean is $E[(\bar{X} - \mu)^2]$. Given the goal of estimating the population mean, we want the mean squared error to be as small as possible.

Recall from Chapter 2 that a weighted mean refers to a situation where each observation is multiplied by some constant and the results are added. Using standard notation, if we observe the values $X_1, \ldots, X_n$, a weighted mean is

$$w_1 X_1 + \cdots + w_n X_n,$$

where the weights $w_1, \ldots, w_n$ are specified constants. Under general conditions, the central limit theorem applies to a wide range of weighted means. That is, as the number of observations increases and if we could repeat an experiment billions of times, we would get fairly good agreement between the plot of weighted means and a normal curve. Assuming that the central limit theorem applies, Laplace showed that under random sampling, the weighted mean that tends to be the most accurate estimate of the population mean is the sample mean where each observation gets the same weight, $1/n$. That is, among the infinitely many weighted means we might choose, the sample mean $\bar{X}$ tends to be most accurate based on its average squared distance from the population mean.

In Chapter 2 we saw that when measuring error, we get a different result when using squared error from when using absolute error. Rather than measure accuracy using mean squared error, suppose we use the average absolute error instead. That is, we measure the typical error made when using the sample mean with $E(|\bar{X} - \mu|)$, the average absolute difference between the sample mean and the population mean. Among all weighted means we might consider, which is optimal? Laplace showed that the usual sample mean is optimal under random sampling. This might seem to contradict our result that when measuring error using absolute values we found that the sample median beat the sample mean, but it does not. As a partial explanation, note that the sample median does *not* belong to the class of weighted means. It involves more than weighting the observations; it requires putting the observations in ascending order. In fact, Laplace found conditions under which the median beats the mean regardless of whether squared error or absolute error is used.

For weighted means, Gauss derived results similar to Laplace, without resorting to the central limit theorem. Using what we now call the *rules of expected values*, Gauss showed that under random sampling the optimal weighted mean for estimating the population mean is the usual sample mean, $\bar{X}$. Put another way, of all the linear combinations of the observations we might consider, the sample mean is most accurate under relatively unrestrictive conditions that allow probability curves to be nonnormal, regardless of the sample size. Again, it should be stressed that this class of weighted means does not include the median or two general classes of estimators, which will be introduced in Chapter 8. Laplace obviously had some concerns about using the sample mean, concerns that were based on theoretical results he derived, but it seems unlikely that he could fully appreciate the seriousness of the problem based on the mathematical tools available to him.

The results by Laplace and Gauss, just described, were actually couched in the much more general framework of multiple regression, which contains the problem of estimating the population mean as a special case. Despite any misgivings Laplace might have had about the least squares principle in general and the mean in particular, he used least squares in his applied work. No doubt this was due in part to the mathematical reality that at the time, more could be done using least squares than with any other method.

THE MEDIAN VERSUS THE MEAN

A special case of what Laplace and Gauss established is that if a probability curve is symmetric about the population mean and we want to find a more accurate estimate of the population mean than the sample mean, under random sampling we must look outside the class of weighted means. One such candidate is the median, and Laplace found that it is more or less accurate than the mean, depending on the equation for the probability curve. If, for example, the equation for the probability curve is the normal curve given by Equation 2.1, the mean beats the median in accuracy. But Laplace was able to describe general conditions under which the reverse is true. This result, when first encountered, is often met with incredulity because the sample median is based on at most the two middle values, with the rest of the observations getting zero weight. A common and natural argument is that because the median throws away most of the data and the sample mean uses all the data, the sample mean must be more accurate. This line of reasoning is erroneous, however, but it is too soon to explain why. For the moment we merely illustrate graphically that the median can beat the mean in some cases but not others.

Figure 4.1 illustrates the sense in which the mean is more accurate than the median when sampling from a normal curve. Twenty observations were generated on a computer from a normal curve having a population mean of 0, and the mean and median were computed. This process was repeated four thousand times and the results plotted. Notice that the tails of the plotted means are closer to the center than the median. That is, both the mean and median differ from 0, the value they are trying to estimate. But the median is more likely to be less than −0.4, or greater than 0.4. That is, the means are more tightly centered around the value 0 than the medians. So the mean is, on average, more accurate.

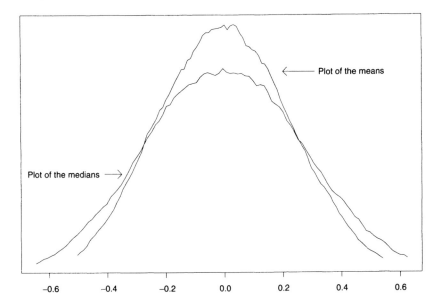

FIGURE 4.1 • When sampling from a normal curve, the sample mean tends to be closer to the center (the population mean) than the median. That is, the sample mean is more likely to provide the more accurate estimate of the population mean.

We repeat our computer experiment, but this time we generate observations from Laplace's probability curve shown in Figure 2.2. The results are shown in Figure 4.2. We see that the median beats the mean; the medians are more tightly centered around the value 0 and are, on average, more accurate. (This result can be verified using a mathematical argument.)

We end this section with three important points. First, if there is an optimal estimator in terms of accuracy for any probability curve we might consider, it has to be the sample mean because nothing beats the mean under normality. We have just seen, therefore, that the perfect estimator does not exist. What, you might ask, is the point of worrying about this? The answer is that in Part II of this book, we will see estimators that perform almost as well as the mean when sampling from a normal probability curve, but for even slight departures from normality, these estimators can be vastly more accurate. Simultaneously, the sample mean never offers as striking an advantage. Put another way, the perfect estimator does not exist, but arguments can be made, based on several criteria, for being highly unsatisfied with the mean in a wide range of realistic situations.

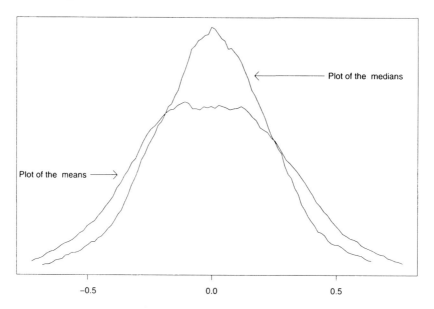

FIGURE 4.2 • When sampling from a Laplace probability curve, the sample median tends to be closer to the population mean than the sample mean. That is, the median tends to be a more or less accurate estimate of the center of a symmetric probability curve, than the mean, depending on the probability curve from which observations are sampled.

Second, it is easy to dismiss the illustration in Figure 4.2 on the grounds that Laplace's probability curve is rather unrealistic compared to what we find in applied work. There is certainly merit to this argument. The only goal here is to illustrate that about two hundred years ago, it was known that the median can beat the mean in accuracy, not only for Laplace's probability curve but for a wide range of situations. More realistic departures from normality are taken up in Chapter 7. Another criticism is that although the median beats the mean, it might seem that the increased accuracy is not that striking. Chapter 7 will describe situations where under arbitrarily small departures from normality, the median is vastly more accurate.

Finally, it cannot be stressed too strongly that we are not leading up to an argument for routinely preferring the median over the mean. There are situations where routine use of the median is a reasonable option, but there are situations where alternatives to both the mean and the median currently seem best. For a wide range of problems, we will see that there are practical concerns with the median yet to be described.

REGRESSION

Next we consider simple regression where we have one predictor (X) and one outcome variable (Y), and the goal is to understand how these measures are related. A common strategy is to fit a straight line to the scatterplot of points available to us. For example, E. Sockett and his colleagues conducted a study with the goal of understanding various aspects of diabetes in children. A specific goal was to determine whether the age of a child at diagnosis could be used to predict a child's C-peptide concentrations. One way of attacking the problem is to try to determine the *average* C-peptide concentrations given a child's age. In particular, we might try a rule that looks like this:

$$\hat{Y} = \beta_1 X + \beta_0,$$

where Y is the C-peptide concentration and X is the child's age. If, for example, we take the slope to be $\beta_1 = .0672$ and the intercept to be 4.15, then given that a child is seven years old, the predicted C-peptide concentration would be

$$\hat{Y} = .0672 \times 7 + 4.15 = 4.62.$$

That is, among all seven-year-old children, the average C-peptide concentration is estimated to be 4.62.

As indicated in Chapter 2, there are many ways the slope (β_1) and intercept (β_0) might be estimated. Among these is the least squares estimator, which is routinely used today. When is this estimator optimal? The first step answering this question hinges on the notion of what mathematicians call a homoscedastic error term. Let's continue to use the diabetes study and temporarily focus on all children who are exactly seven years old. Of course, there will be some variation among their C-peptide concentrations. If all seven-year-old children could be measured, then we would know the population mean for their C-peptide concentrations and the corresponding variance. That is, we would know both the population mean (μ) and the population standard deviation (σ) for this particular group of children. (In more formal terms, we would know the conditional mean and the conditional variance of C-peptide concentrations given that a child's age is 7.) In a similar manner, if we could measure all children who are exactly 7.5 years old, we would know their population mean and standard deviation. *Homoscedasticity* refers to a situation where, regardless of the age of the children, the population standard deviation is always the same. So if we are told that the standard deviation of C-peptide concentrations for eight-year-old children is 2, homoscedasticity implies that the standard deviation among children who are 5, 6, 7, or any age we might pick, is again 2.

To graphically illustrate homoscedasticity, assume that for any X value we might pick, the average of all Y values, if only they could be measured, is equal to $X + 10$. That is, the straight line in Figure 4.3 could be used to determine the mean of Y given a value for X. So if we obtain a large number of Y values corresponding to $X = 12$, say, their average would be $X + 12 = 2 + 12 = 14$, and their values, when plotted, might appear as shown in the left portion of Figure 4.3. In a similar manner, a large number of Y values where $X = 20$ would have mean 22 and might appear as plotted in the middle portion of Figure 4.3. Notice that for each of the X values considered, the spread or variation among the corresponding Y values is the same. If this is true for any X value we pick, not just the three shown here, then we have a homoscedastic model.

Heteroscedasticity refers to a situation where the variances are not homoscedastic. For example, if infinitely many measurements could be made, a plot of the Y values might appear as shown in Figure 4.4. The variation among the Y values at $X = 12$ is greater than the variation at $X = 20$. Moreover, the variation at $X = 30$ differs from the variation at $X = 12$ and $X = 20$.

With the notion of homoscedasticity in hand, we can now describe a major result derived by Gauss and commonly referred to as the Gauss-Markov theorem. Chapter 2 noted that the least squares estimator of the slope and intercept can be viewed as a weighted mean of the Y values. Gauss showed that if there is homoscedasticity, then among all the weighted means we might consider, the weights corresponding to the least squares estimator are optimal in terms of minimizing the expected squared error. That is, among all linear combinations of the Y values, least squares minimizes, for example, $E(b_1 - \beta_1)^2$, the expected squared difference between the estimated slope (b_1) and its true value (β_1).

Gauss did more. He also derived the optimal estimator under a heteroscedastic model, meaning that the variation among the Y values changes with X. Again attention is restricted to estimators that are weighted means of

REGRESSION

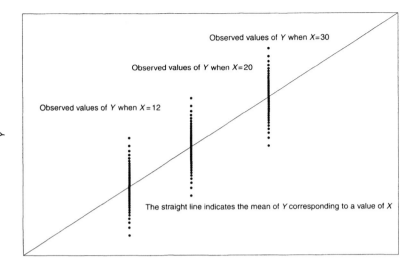

FIGURE 4.3 • In regression, homoscedasticity refers to a situation where the variation of the Y values does not depend on X. Shown are plotted values of Y corresponding to X equal to 12, 20, and 30. For each X value, the variation among the Y values is the same. If this is true for all values of X, not just the three values shown here, there is homoscedasticity.

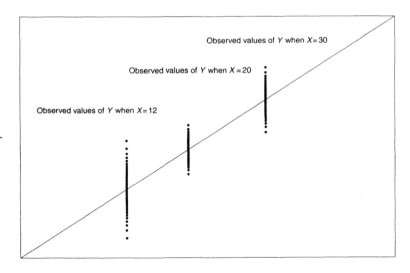

FIGURE 4.4 • Heteroscedasticity refers to a situation where the variation of Y changes with X. Here there is more variation among the Y values, given that $X = 12$, than in the situation where $X = 20$, so there is heteroscedasticity.

the Y values. The optimal weights turn out to depend on the *unknown* variances. That is, if the variance of the Y values corresponding to any X were known, then optimal weights for estimating the slope could be determined even when the variances do not have a common value. Moreover, Gauss derived his result in the general context where there are multiple, not just one, predictors for estimating Y.

But how do we implement Gauss's result if there is heteroscedasticity? If we could estimate how the variance of Y changes with X, we would have a possible solution; in some situations this can be done. But for a wide range of situations, until fairly recently, no satisfactory method was available for implementing this strategy. So by tradition, starting with Gauss and Laplace, there are many situations where applied researchers typically use the least squares estimator, effectively assuming that there is homoscedasticity. For one common goal, to be described later, this assumption is satisfactory, but in terms of accuracy when estimating the slope and intercept, this strategy can be disastrous. One reason is that in some situations, using the optimal weights can result in an estimate of the slope and intercept that is hundreds, even thousands of times more accurate! Naturally, one would like to know the optimal weights for these cases, or at least be able to find a reasonably accurate estimate of them. Another, perhaps more common, problem is described in the next section.

CONFIDENCE INTERVALS

In the year 1811, at the age of 62, Laplace made another major breakthrough; he developed what we now call the frequentist approach to problems in statistics. We cannot develop a full appreciation of his accomplishment here, but it is noted that in 1814, Laplace used his new approach to derive a new solution to a problem of extreme importance: He found a new approach to computing what is called a *confidence interval*. Today, his method is routinely covered in every introductory statistics course. Although the notion of a confidence interval was not new, the method derived by Laplace marks one of the greatest milestones in statistical theory.

To explain the idea behind Laplace's approach, we return to the problem of estimating the population mean. For illustrative purposes, we again consider the problem of estimating the average cholesterol levels of some population of individuals. Imagine that based on twenty individuals the average cholesterol level (the sample mean, $\bar{X}$) is 225. Can we be reasonably certain that the unknown population mean is not 240 or 260? Can we, with a reasonable degree of certainty, rule out the values 200 or 180? How close is the sample mean to the population mean?

A solution can be derived if we can somehow approximate the probabilities associated with values we might get for the sample mean. This means we need a method for approximating the probability that the sample mean will be less than 180, or less than 200, or less than c for any constant c we might pick. Laplace's method consists of two components. First, determine an expression for the variance of infinitely many sample means if we could somehow repeat our experiment ad infinitum. Even though we cannot repeat an experiment infinitely many times, it turns out that an expression for this variation can be derived and estimated from the observations we make. Second, assume that a plot of the infinitely many sample means, if only they could be observed, follows a normal curve. This latter assumption is reasonable if the individual observations arise from a normal curve or if a sufficiently large number of observations is used to compute the sample mean, in which case we can invoke the central limit theorem. As noted in Chapter 3, the expression *sufficiently large* is rather vague, but for the moment we ignore this issue and focus on deriving an expression for the variation among sample means.

Let's look at the first step for applying Laplace's method. It can be shown that if we randomly sample some observations and compute the sample mean, and if we repeat this process infinitely many times yielding infinitely many sample means, the average of these sample means will be equal to μ, the population mean. This is typically written more succinctly as $E(\bar{X}) = \mu$, in which case the sample mean is said to be an *unbiased* estimate of the population mean. (On average, it gives the correct answer.)

The variance of the sample mean refers to the variation among infinitely many sample means using a measure that is essentially the same as the sample variance, s^2. For example, imagine we repeat an experiment many times and observe the sample means 36, 89, ..., 42, and 87, and suppose the average of these sample means, typically called a *grand mean*, is 53. We could then compute the squared distance of each sample mean from 53 and average the results. This would provide an estimate of $\text{VAR}(\bar{X})$, the variance of infinitely many sample means. The variance of the sample mean is typically called the *squared standard error* of the sample mean. But we just saw that if we average infinitely many sample means, we get the population mean. So it turns out that the variance of the sample mean is the expected squared distance between the sample mean and the population mean. That is,

$$\text{VAR}(\bar{X}) = E(\bar{X} - \mu)^2.$$

For the moment, we simply remark that under random sampling, this last equation can be shown to reduce to

$$\text{VAR}(\bar{X}) = \frac{\sigma^2}{n}. \tag{4.1}$$

That is, the variance of the sample means is just the population variance associated with the probability curve of the individual observations divided by the sample size. As we have seen, the population variance can be estimated with the sample variance, s^2, so we can also estimate the variance of the sample mean, even without repeating our experiment: We simply divide s^2 by the number of observations, n. A very important point, however, one not typically stressed in an introductory applied course, is that the result in Equation 4.1 depends in a crucial way on the independence of the observations. Some understanding of this point is vital if we want to avoid an erroneous strategy discussed in Part II of this book.

It might help to elaborate a little about independence. Imagine you conduct a weight-loss study and the first individual loses 15 pounds. Now, there is some probability that the second individual in your study will lose less than 5 pounds, or less than 10 pounds, or less than c pounds for any constant c we pick. If any of these probabilities are altered by knowing that the first subject lost 15 pounds, then the results for these two individuals are said to be *dependent*. If the probabilities are not altered, we have *independence*. By definition, a *random sample* means that all observations are independent. Under independence, the expression for the variation among sample means, given by Equation 4.1, holds. But if there is dependence, in general, Equation 4.1 is no longer valid. (There is an exception, but it is not important here.)

CONFIDENCE INTERVAL FOR THE POPULATION MEAN

Recall that for any normal distribution, the probability of being within 1.96 standard deviations of the mean is exactly .95. This fact can be used to derive the following result. If we assume $\bar{X}$ has a normal distribution with mean μ and variance σ^2/n, then there is a .95 probability that the interval

$$\left(\bar{X} - 1.96\frac{\sigma}{\sqrt{n}}, \bar{X} + 1.96\frac{\sigma}{\sqrt{n}} \right) \tag{4.2}$$

contains the *unknown* population mean. This means that if we were to repeat our experiment infinitely many times, each time using n values to compute the

CONFIDENCE INTERVAL FOR THE POPULATION MEAN

mean, then 95 percent of the confidence intervals, computed in the manner just described, would contain μ. For this reason, Equation 4.2 is called a *.95 confidence interval* for μ. When we use the sample mean to estimate the unknown population mean, the length of the confidence interval gives us some sense of how accurate this estimate is.

Notice that the smaller $\sigma/\sqrt{n}$ happens to be, the shorter the length of the confidence interval given by Equation 4.2 is. Put another way, the variance of the sample mean (σ^2/n), which is also its mean squared error, measures how well the sample mean estimates the population mean. Thus, a small variance for the sample mean means that we are better able to narrow the range of reasonable values for the population mean based on the observations we make. That is, we get a shorter confidence interval. We can shorten the confidence interval by increasing the sample size, but the population variance is beyond our control. However, it is the connection between the variance and the length of the confidence interval that turns out to be crucial for understanding modern insights covered in subsequent chapters.

Typically, the population variance is not known, so Laplace estimated it with the sample variance. That is, he used

$$\left(\bar{X} - 1.96 \frac{s}{\sqrt{n}},\ \bar{X} + 1.96 \frac{s}{\sqrt{n}} \right) \tag{4.3}$$

as an approximate .95 confidence interval and assumed that the sample size was large enough so that the actual probability coverage would be approximately .95. (Actually, Laplace routinely used 3 rather than 1.96 in Equation 4.3, with the goal of attaining a probability coverage higher than .95.)

A brief illustration might help. The following nineteen values ($n = 19$) are from a study dealing with self-awareness:

77 87 87 114 151 210 219 246 253 262 296 299 306 376 428 515 666 1310 2611.

The sample mean is $\bar{X} = 448$, and the sample standard deviation is $s = 594.66$, so Laplace's .95 confidence interval for the population mean is (180.6, 715.4). That is, assuming normality and that the sample variance provides a tolerably accurate estimate of the population variance, we can be reasonably certain that the population mean is somewhere between 180.6 and 715.4. In this particular case the length of the confidence interval is rather large because the sample standard deviation is large. A relatively large standard deviation was to be expected because based on the boxplot in Chapter 3, the values

1310 and 2611 are outliers, and as already noted, outliers can inflate the sample standard deviation. One way of getting a shorter confidence interval is to increase n, the sample size.

It is noted that rather than compute a .95 confidence interval, one could just as easily compute a .9 or .99 confidence interval. The desired probability coverage is often set to .95, but any probability could be used in principle. Here, primarily for convenience, attention is focused on .95.

When we apply Laplace's method, an obvious concern is that in reality observations do not follow a normal curve and the population variance is not known but being estimated. So a practical issue is the accuracy of any confidence interval we compute. In the illustration, we computed a confidence interval that we hope has probability coverage .95, but in reality its probability coverage is not .95 because our assumptions are only approximations of reality. The central limit theorem says that with a large enough sample, these concerns can be ignored. But when we are given some data and asked to compute a confidence interval, how do we know whether a sample is large enough to ensure that the probability coverage is reasonably close to .95? This is an extremely difficult problem that we will return to at various points in this book.

CONFIDENCE INTERVAL FOR THE SLOPE

For many important applied problems, some approaches are mathematically intractable, or at best very difficult to implement. So naturally, even if these methods have superior theoretical properties, they cannot be employed. In practical terms, do the best you can with the tools you have. To begin to foster an appreciation for how this has shaped modern practice, a general form of Laplace's method for computing a confidence interval is given here and illustrated when dealing with the slope of a regression line.

Laplace's frequentist approach to computing confidence intervals generalizes to a wide range of situations in a manner that looks something like this. We have some unknown number, called a *parameter*, that reflects some characteristic of a population of subjects. Let's call it θ where θ might be the population mean, median, slope of a regression line, or any other parameter of interest. Let $\hat{\theta}$ be some estimate of θ we get from our data. If, for example, θ represents the population mean, then $\hat{\theta}$ could be the sample mean, $\bar{X}$. If θ is the slope of a regression line, then $\hat{\theta}$ could be the least squares regression estimate of the slope. Suppose we can derive an expression for $\text{VAR}(\hat{\theta})$, the squared standard error of $\hat{\theta}$. That is, VAR($\hat{\theta}$) represents the variation among infinitely many $\hat{\theta}$ values if we could repeat a study infinitely many times. For the sample mean, VAR($\hat{\theta}$) is σ^2/n, as already noted. Of course, this means that we have an expression for the standard error as well; it is simply the square root of the squared standard error. Then applying the central limit theorem, in effect assuming that a plot of the infinitely many $\hat{\theta}$ values is normal, a .95 confidence interval for θ is given by

$$\left(\hat{\theta} - 1.96\text{SE}(\hat{\theta}),\ \hat{\theta} + 1.96\text{SE}(\hat{\theta})\right), \tag{4.4}$$

where $\text{SE}(\hat{\theta}) = \sqrt{\text{VAR}(\hat{\theta})}$ is the standard error. This is a generalization of Equation 4.2. The main point is that Equation 4.4 can be employed given an expression for the standard error and an estimate of the standard error based on observations we make.

Laplace was able to apply his method to the problem of computing a confidence interval for the slope of a regression line, but to implement the method he assumed homogeneity of variance. Part of the reason is that if we use least squares to estimate the slope, homogeneity of variance yields an expression for the standard error of the least squares estimator that can be estimated based on observations we make. Without this assumption, estimating the standard error becomes very difficult for many situations that arise in practice.

To see why, consider again the diabetes study. If we have twenty children who are exactly age 7, we can estimate the variance of the corresponding C-peptide levels with the sample variance. But in reality we might have only one child who is exactly 7—the others might be age 7.1, 7.3, 7.5, 6.2, 5.1, and so on. With only one child who is exactly 7, we have only one observation for estimating the variance of the C-peptide levels among children age 7, meaning that it cannot be estimated because we need at least two observations to compute a sample variance. Even if we have two children who are exactly 7, estimating the variance based on only two observations is likely to yield an inaccurate result. By assuming the variance of C-peptide levels does not change with age, we circumvent this technical problem and end up with a method that can be applied. In particular, we are able to use all of the data to estimate the assumed common variance. Put another way, to avoid mathematical difficulties, we make a convenient assumption and hope that it yields reasonably accurate results.

One might argue that for children close to age 7, we can, for all practical purposes, assume homoscedasticity. For example, we might use all children between the ages of 6.8 and 7.2 to estimate the variance of C-peptide levels among children who are exactly 7. But how wide of an interval can and should we use? Why not include all children between the ages of 6.5 and 7.5? Methods for dealing with this issue have been derived, some of which have considerable practical value, but the details are deferred for now.

It turns out that there is a situation where employing a method based on the assumption of homogeneity of variance is reasonable: when the two measures are independent. If C-peptide levels are independent of age, it follows that the variance of the C-peptide levels does not vary with age. Moreover, the slope of the regression line between age and C-peptide levels is exactly zero. So if we want to establish dependence between these two measures we can accomplish our goal by assuming independence and then determining whether our data contradicts our assumption. If it does, we conclude that our assumption of independence is incorrect. That is, we conclude that the two variables of interest are dependent. As we shall see in Part II, a problem with this approach is that by assuming homoscedasticity, we might mask an association that would be detected by an approach that permits heteroscedasticity.

For future reference, the details of Laplace's method for the slope are outlined here for the special case where there is homoscedasticity. As suggested by Gauss, and as is typically done today, an estimate of the assumed common variance is obtained in the following manner. First compute the least squares estimate of the slope and intercept, then compute the corresponding residuals. Next, square each residual, sum the results, and finally divide by $n - 2$, the number of pairs of observation minus 2. The resulting value estimates the assumed common variance, which we label $\hat{\sigma}^2$. The squared standard error of the least squares estimate of the slope is estimated to be $\hat{\sigma}^2/[(n-1)s_x^2]$, where s_x^2 is the sample variance of the X values. Letting b_1 be the estimate of the slope, an approximate .95 confidence interval for the true slope (β_1) is

$$\left(b_1 - 1.96\frac{\hat{\sigma}}{\sqrt{(n-1)s_x^2}},\ b_1 + 1.96\frac{\hat{\sigma}}{\sqrt{(n-1)s_x^2}}\right). \quad (4.5)$$

As a brief illustration, let's compute Laplace's confidence interval using Boscovich's data in Table 2.1. The least squares estimates of the slope and intercept are 723.44 and 56,737.4, respectively. The resulting residuals are

$$13.57393,\ 83.48236,\ -94.68105,\ -80.27228,\ 77.89704. \quad (4.6)$$

Squaring and summing the residuals yields 28,629.64, so the estimate of the assumed common variance is $\hat{\sigma}^2 = 28,629.64/3 = 9543.2$. Dividing by $(n-1)s_x^2 = 0.391519$ yields 24,374.85. Finally, the .95 confidence interval is

(417.4, 1029.4). Assuming normality, homoscedasticity, and that an accurate estimate of the variance has been obtained, there is a .95 probability that this interval contains the true slope.

If in reality the Y values (or the residuals) have a normal distribution, and if Y and X are independent, this confidence interval is exact: The probability coverage will be precisely .95. (Independence between X and Y implies homoscedasticity.) But if X and Y are dependent, there is no compelling reason to assume homoscedasticity. And if there is heteroscedasticity, the confidence interval just described can be extremely inaccurate—even under normality. For example, the actual probability coverage might be less than .5. That is, we might claim that with probability .95 our computed confidence interval contains the true value for the slope, but in reality the probability might be less than .5. In terms of Type I errors, if we test at the .05 level, the actual probability of a Type I error might exceed .5!

A reasonable suggestion is to check for homoscedasticity, and if it seems to be a good approximation of reality, use Equation 4.5 to compute a confidence interval. The good news is that there are several methods for testing this assumption. Unfortunately, it is unclear when these methods are sensitive enough to detect situations where the assumption should be abandoned. In fact, all indications are that these methods frequently fail to detect situations where the assumption of homoscedasticity is violated and leads to erroneous results. So, at least for the moment, it seems we are better off using a method that performs relatively well when there is homoscedasticity and that continues to perform well in situations where there is heteroscedasticity. Today, such methods are available, as we shall see. Moreover, these more modern methods can give substantially different results. That is, it is not academic which method is used.

A SUMMARY OF KEY POINTS

- The Gauss-Markov theorem says that among all the weighted means we might consider for estimating μ, the sample mean is optimal in terms of mean squared error. But there are estimators outside the class of weighted means that can be substantially more accurate than the sample mean. One example is the median when sampling from a heavy-tailed distribution. (Other examples are described in Part II of this book.)

- The least squares estimate of the slope of a regression line, typically used in applied work, assumes homoscedasticity. When this assumption is violated, alternate estimators (described in Part II) can be substantially more accurate, even under normality.

- Laplace's strategy for computing a confidence interval was introduced. It forms the foundation of methods typically used today. The method was illustrated when there is interest in μ or the slope of the least squares regression line. In the latter case, the method can be extremely inaccurate under heteroscedasticity, even when sampling from a normal distribution. Practical problems arise when using Equation 4.3 to compute a confidence interval for μ. (Details will be covered in Chapter 5.)

CHAPTER 5

HYPOTHESIS TESTING AND SMALL SAMPLE SIZES

One of the biggest breakthroughs in the past forty years is the derivation of inferential methods that perform well when sample sizes are small. Indeed, some practical problems that seemed insurmountable only a few years ago have been solved. But to appreciate this remarkable achievement, we must first describe the shortcomings of conventional techniques developed during the first half of the twentieth century—methods that are routinely used today. At one time it was generally thought that these standard methods are insensitive to violations of assumptions, but a more accurate statement is that they seem to perform reasonably well (in terms of Type I errors) when groups have identical probability curves or when performing regression with variables that are independent. If, for example, we compare groups that happen to have different probability curves, extremely serious problems can arise. Perhaps the most striking problem is described in Chapter 7, but the problems described here are also very serious and are certainly relevant to applied work.

HYPOTHESIS TESTING

We begin by summarizing the basics of hypothesis testing, a very common framework for conveying statistical inferences based on observations we make. Some of Laplace's applied work is the prototype for this approach, and it was formally studied and developed during the early twentieth century. Of particular note is the work by Jerzey Neyman, Egon Pearson (son of Karl Pearson, whom we met in Chapter 1), and Sir Ronald Fisher.

For the case of the population mean corresponding to a single population of individuals, hypothesis testing goes like this. There is a desire to determine whether one can empirically rule out some value for the population mean. For example, after years of experience with thousands of individuals, a standard treatment for migraine headache might have an average effectiveness rating of 35 on some scale routinely used. A new method is being considered, so there is the issue of whether it is more or less effective than the standard technique. Temporarily assume that it is just as effective or worse on average. That is, we assume that the unknown average effectiveness of the new procedure is less than or equal to 35. This is written as

$$H_0 : \mu \leq 35,$$

and is an example of what is called a *null hypothesis*. If there is empirical evidence that this null hypothesis is unreasonable, we reject it in favor of the alternative hypothesis

$$H_1 : \mu > 35.$$

That is, we conclude that the null hypothesis is unlikely, in which case we decide that the new treatment has a higher effectiveness rating, on average.

Assume that sampling is from a normal curve and the standard deviation is known. If $\mu = 35$, meaning that the null hypothesis is true, then it can be shown that

$$Z = \frac{\bar{X} - 35}{\sigma/\sqrt{n}}$$

has a standard normal distribution. Notice that Z is greater than zero if the sample mean is greater than 35. The issue is, given some observations yielding a sample mean, how much greater than 35 must the sample mean be to rule out the conjecture that the population mean is less than or equal to 35? Alternatively, by how much must Z exceed 0 to reject H_0? If we reject when $Z > 1.645$, properties of the normal curve indicate that the probability of rejecting, when the null hypothesis is actually true, is .05. If we reject when $Z > 1.96$, this probability is .025, again assuming normality.

Suppose we reject when Z exceeds 1.645. If the null hypothesis is true, by chance we might erroneously reject. That is, even if $\mu = 35$, there is some possibility that we will get a value for Z that is bigger than 1.645. Deciding the null hypothesis is false, when in fact it is true, is called a *Type I error*. The probability of a Type I error is typically labeled α. Here, still assuming normality, the probability of a Type I error is $\alpha = .05$.

The computational details are not particularly important for our current purposes, so for brevity an illustration is not given. (Most introductory books give detailed illustrations.) However, there are two slight variations that should be mentioned. They are:

$$H_0 : \mu \geq 35$$

and

$$H_0 : \mu = 35.$$

In the first, you start with the hypothesis that the mean is greater than or equal to 35, and the goal is to determine whether Z is small enough to reject. In the second, a specific value for μ is hypothesized and you reject if Z is sufficiently small or sufficiently large. Rejecting the null hypothesis means that we rule out the possibility that the population mean is 35. If, for example, you reject when $|Z| > 1.96$, it can be shown that the probability of a Type I error is $\alpha = .05$ when sampling from a normal curve.

The method for testing $H_0 : \mu = 35$ is closely connected to the .95 confidence interval previously described. If the confidence interval for μ does not contain the value 35, a natural rule is to reject $H_0 : \mu = 35$. If we follow this rule, the probability of a Type I error is $1 - .95 = .05$, again assuming normality. If we compute a .99 confidence interval instead, the probability of a Type I error is $1 - .99 = .01$.

A *Type II error* is failing to reject when in fact the null hypothesis is false. The probability of a Type II error is often labeled β (which should not be confused with the regression parameters β_1 and β_0). *Power* is the probability of rejecting when in fact the null hypothesis is false. Power is the probability of coming to the correct conclusion when some speculation about the mean (or some other parameter of interest) is false. In our illustration, if the population mean is 40 and we are testing the hypothesis that it is less than or equal to 35, power is the probability of rejecting. If, for example, the probability of a Type II error is $\beta = .7$, power is $1 - \beta = 1 - .7 = .3$. That is, there is a 30 percent chance of correctly rejecting.

In our illustration, a Type I error is a concern because we do not want to recommend the new method for treating migraine if it actually has no value. Simultaneously, we do not want to commit a Type II error. That is, if the new treatment is more effective, failing to detect this is obviously undesirable. So, anything we can do to increase power is of interest.

Table 5.1 summarizes the four possible outcomes when testing hypotheses. As already indicated, two outcomes represent errors. The other two reflect correct decisions about the value of μ based on the data at hand. In our illustration we tested the hypothesis that the mean is less than 35, but of course, there is nothing special about the value 35. We could just as easily test the hypothesis that the population mean is 2, 60, or any relevant value.

TABLE 5.1 • FOUR POSSIBLE OUTCOMES WHEN TESTING HYPOTHESES

	Reality	
Decision	H_0 True	H_0 False
H_0 True	Correct decision	Type II error
H_0 False	Type I error	Correct decision

Assuming normality, there are methods for assessing power and the probability of a Type II error (determining the value for β), given the sample size (n), the Type I error value chosen by the investigator (α), and the difference between the actual and hypothesized values for the mean. Some introductory books give details, but they are not relevant here. What is more important is understanding the factors that affect power. One of these factors plays a central role in subsequent chapters.

Consider the hypothesis $H_0 : \mu = \mu_0$, where μ_0 is some constant of interest. In our migraine illustration, $\mu_0 = 35$. Power—our ability to detect situations where the null hypothesis is false—is related to n, α, σ, and $\mu - \mu_0$ in the following manner:

- As the sample size, n, gets large, power goes up, so the probability of a Type II error goes down.

- As α goes down, in which case the probability of a Type I error goes down, power goes down and the probability of a Type II error goes up. That is, the smaller α happens to be, the less likely we are to reject when in fact we should reject because (unknown to us) the null hypothesis is false.

- As the standard deviation, σ, goes up, with n, α and $\mu - \mu_0$ fixed, power goes down.

- As $\mu - \mu_0$ gets large, power goes up. (There are exceptions, however, which are described later in this chapter.)

Notice that Type I errors and power are at odds with one another. You can increase power by increasing α, but then you also increase the probability of a Type I error.

Assessing power is complicated by the fact that we do not know the value of the population mean. So we must play the "what if" game. That is, we ask ourselves what if $\mu = 40$ or what if $\mu = 45$? How much power is there? Power will vary depending on how far the unknown value of the population mean happens to be from its hypothesized value (μ_0).

If we hypothesize that $\mu = 35$, and in reality the mean is 40, we want to reject because the null hypothesis is false. But if the mean is really 60, we want power to be even higher because the difference between the hypothesized value and the actual value is greater. More generally, we want power to increase when $\mu - \mu_0$ increases because we are moving away from the null hypothesis. (Hypothesis testing methods that satisfy this criterion are said to be *unbiased*.) As we shall see, there are commonly used techniques where this property is not always achieved. That is, among the previously listed features that affect power, the last one is something we desire, but it is not always achieved.

Of particular importance in this book is how the variance is related to power. Again imagine that we want to test the hypothesis that the population mean is less than or equal to 35. For illustrative purposes, assume the sample size is 16, the population standard deviation is 20, and we want the probability of a Type I error to be $\alpha = .05$. It can be seen that if the population mean is actually 41, power (the probability of rejecting) is .148. That is, there is a 14.8 percent chance of correctly deciding that the hypothesized value for the population mean is false. But if the standard deviation is 10 instead, power is .328. This illustrates that as the population standard deviation goes down, power goes up.

The connection between variance and power is so important that we will describe it again from a slightly different perspective to make sure it is clear. Recall that the accuracy of the sample mean (its mean squared error) is measured by the variance of the sample mean, which is σ^2/n. The smaller σ^2/n happens to be, the more likely the sample mean will be close to the value it is estimating, the population mean. In particular, the confidence interval for the population mean gets shorter as σ^2/n gets smaller, a fact that was mentioned in Chapter 4 in connection with Equation 4.2. Imagine, for example, that the population mean is 41 and we are testing the hypothesis that the population mean is 35. Then the smaller σ^2/n happens to be, the more likely we are to get a confidence interval that does not contain 35. That is, we are more likely to reject $H_0: \mu = 35$. We see that increasing the sample size increases power because it lowers the value of σ^2/n. Yet it is the connection between the population variance and power that will prove to be most important here. For now, the key point is that power will be low if the population variance happens to be large.

Figure 5.1 graphically provides a rough indication of why power goes up as the standard deviation goes down. Again imagine that unknown to us, the population mean is 41. Figure 5.1 shows the probability curves for the sample mean when $\sigma = 20$ versus $\sigma = 10$, when sampling from a normal distribution. As expected, based on properties of the normal curve already covered, the sample means are more tightly clustered around the population mean for the smaller standard deviation. Notice that for the case $\sigma = 20$, the probability of the sample mean being less than the hypothesized value, 35, is higher than $\sigma = 10$. (The area under the curve and below the point 35 is higher for $\sigma = 20$ than for $\sigma = 10$.) Of course, if we get a sample mean that is less than 35, this is consistent with the hypothesis that the population mean is less than 35, so the null hypothesis would not be rejected. This corresponds to a Type II error because the null hypothesis is false. This suggests that we are more likely to reject when $\sigma = 10$, and a more careful mathematical analysis verifies that this is the case.

THE ONE-SAMPLE T TEST

Now we consider the problem of computing confidence intervals and testing hypotheses when the unknown population variance is estimated with the sample variance. As noted in Chapter 4, Laplace derived a method for computing confidence intervals that assumes the variance is known. In his applied work, Laplace simply estimated the population standard deviation (σ) with the sample standard deviation (s) and appealed to his central limit theorem. In essence, the method assumes that if we subtract the population mean from the sample mean, and then divide by the estimated standard error of the sample mean, we get a standard normal curve (with mean 0 and variance 1). That

THE ONE-SAMPLE T TEST

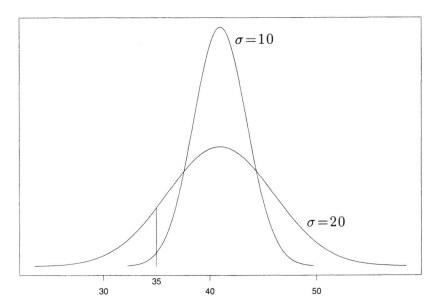

FIGURE 5.1 • Shown are the sampling distributions of the sample mean when $\sigma = 10$ versus $\sigma = 20$ and sampling is from a normal curve. For the smaller standard deviation ($\sigma = 10$), the sample mean is more tightly distributed about the population, $\mu = 41$. So the sample mean is more likely to be greater than the hypothesized value for the population mean, 35, than the situation where $\sigma = 20$. That is, power will be greater when the population standard deviation is small.

is, the method assumes that

$$T = \frac{\bar{X} - \mu}{s/\sqrt{n}} \quad (5.1)$$

has a standard normal distribution and the central limit says that with enough observations, reasonably accurate probability coverage will be obtained. In a similar manner, we can test hypotheses about the population mean when the standard deviation is not known, and we can control the probability of a Type I error provided the sample size is large enough that T has a standard normal distribution.

But this assumption will be incorrect when sample sizes are small, even when observations are sampled from a normal curve. So there are at least two issues: determining how large the sample size must be to achieve reasonably accurate probability coverage when computing a confidence interval and deriving better techniques for situations where Laplace's method gives unsatisfactory results. Chapter 3 noted that when sampling from a light-tailed curve, meaning outliers are relatively rare, the probability curve for the sample mean is approximately normal even when the sample size is fairly small. This might suggest that when using T, nonnormality is not a problem when outliers are rare, but this speculation turns out to be incorrect, as we shall see in the next section.

The first attempt at improving Laplace's method was made by William Gosset during the early years of the twentieth century. Using tables of random numbers to generate observations from normal curves, Gosset generated a small number of observations and computed the mean and standard deviation of these observations and the value of T given by Equation 5.1. He then repeated this process many times, yielding a collection of T values, and he used the results to derive an approximation of the probability curve associated with T. In 1908 he published a mathematical derivation of his curve. That is, he derived a mathematical method for determining the probability that T is less than any constant we might choose, assuming normality. More specifically, given any number c, $P(T < c)$ can be determined, and the answer depends only on the sample size used to compute the sample mean. For large sample sizes, the probability curve associated with the T values becomes indistinguishable from the (standard) normal curve, a result predicted by the central limit theorem. Gosset worked as a chemist for a brewery and was not immediately allowed to publish his results. When he published his mathematical derivation of the probability curve for T, he did so under the pseudonym Student. The resulting probability curve describing the plot of T values is known today as Student's T distribution.

Figure 5.2 shows the probability curve associated with T when five observations are used to compute the mean and variance and sampling is from a normal curve. As is evident, it is bell-shaped and centered around zero. What is less evident is that this probability curve does not belong to the family of normal curves, even though it is bell-shaped. Normal curves are bell-shaped, but there are infinitely many bell-shaped curves that are not normal. The key point is that whenever Student's T distribution is used, it is assumed that plots of infinitely many T values, if they could be obtained, would appear as in Figure 5.2.

Not long after Gosset obtained his result, Sir Ronald Fisher gave a more formal mathematical derivation of the probability curve for T. The complete details are not important here, but there is one aspect of the derivation that is worth mentioning. Recall that when computing the sample variance you begin

THE ONE-SAMPLE T TEST

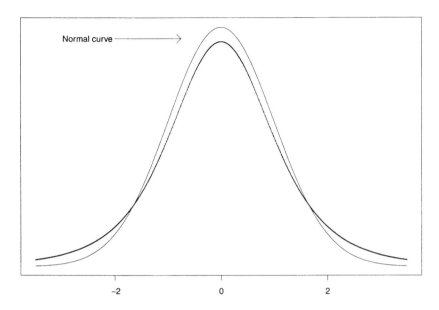

FIGURE 5.2 • The standard normal curve and the probability curve for T when the sample size is five and observations are randomly sampled from a normal curve. As the sample size increases, the probability curve for T looks more like the normal curve. Whenever conventional hypothesis testing methods for means are employed, it is assumed that the probability curve for T is bell-shaped and symmetric about zero, as shown here.

by subtracting the sample mean from each observation, squaring the results, and then adding them. Finally you divide by the sample size (n) minus one. That is, the sample variance is

$$s^2 = \frac{1}{n-1}[(X_1 - \bar{X})^2 + \cdots + (X_n - \bar{X})^2].$$

Because the mean plays a role in determining the sample variance, a natural speculation is that the sample variance (s^2) and sample mean ($\bar{X}$) are dependent. That is, if we are told that the value of the sample mean is 1, this might affect the probabilities associated with sample variance versus not knowing what the mean is. Put another way, given that the sample mean is 1, there is a certain probability that the sample variance will be less than 5. *Independence* between the sample mean and variance implies that this probability will be the same when the sample mean is 2, 3, or any value we choose. *Dependence* refers to situations where this probability does not remain the same. That is,

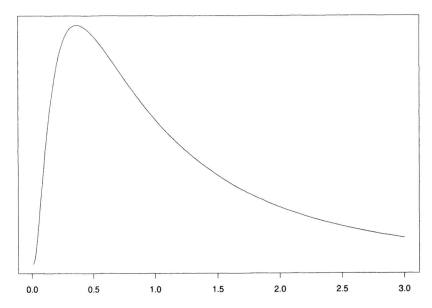

FIGURE 5.3 • An example of a skewed, light-tailed probability curve that is frequently used to check the small-sample properties of statistical methods. Called a lognormal distribution, this particular curve is relatively light-tailed, meaning that when observations are randomly sampled, outliers are relatively rare compared to other curves that are frequently used. The shape of the curve reflects what is commonly found in many applied settings.

the probability that s^2 is less than 5, might be altered if we are told the value of the sample mean. Generally, the sample mean and variance are dependent, but an exception is when observations follow a normal curve.

To add perspective, it might help to graphically illustrate the dependence between the sample mean and sample variance. To do this, twenty observations were generated from the curve shown in Figure 5.3. (The probability curve in Figure 5.3 is an example of a lognormal distribution whose mean is approximately equal to 1.65 and a standard deviation of approximately 2.16. The shape of this curve reflects what we find in many applied problems.) This process was repeated a thousand times, and the resulting pairs of means and variances appear as shown in Figure 5.4. Notice that for sample means with values close to 1, there is little variation among the corresponding sample variances, and all of these variances are relatively small. But for large sample means there is more variation among the sample variances, and the sample

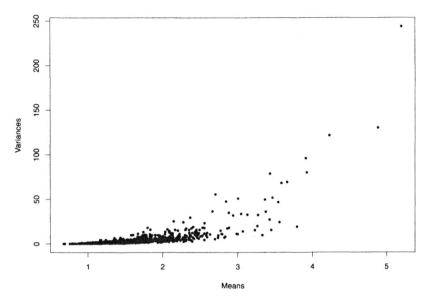

FIGURE 5.4 • Generally, the sample mean and variance are dependent. This is illustrated here when sampling from the curve shown in Figure 5.3. If the sample mean and variance were independent and if we plotted them, we should get a cloud of points with no discernible pattern.

variances tend to be larger than in situations where the sample mean is close to 1. That is, the sample means and variances are dependent.

From a mathematical point of view, the independence between the sample mean and variance, under normality, makes the assumption of normality extremely convenient. Without this independence, the derivation of the probability curve for T is much more complex. So the strategy being suggested by Gosset is to make a simplifying assumption (normality), exploit this assumption to derive the probability curve for T, and then hope that it provides a good approximation for situations that arise in practice. Under normality, Student's (Gosset's) T distribution makes it possible to compute exact confidence intervals for the population mean, and one gains exact control over the probability of a Type I error when testing hypotheses, regardless of how small the sample size is. For a while it seemed to give reasonably accurate results even when sampling from nonnormal distributions, but serious exceptions are now known to exist.

SOME PRACTICAL PROBLEMS WITH STUDENT'S T

We saw in Chapter 3 that when sampling from a skewed, heavy-tailed probability curve, we might need about one hundred observations for the central limit theorem to give accurate results. It turns out that when we take into account that the unknown standard deviation is being estimated, new problems arise, some of which are described here.

Again consider the probability curve shown in Figure 5.3. If we sample observations from this curve, will the resulting probability curve for T be well approximated by the curve we get when sampling from a normal distribution instead? To find out, let's generate twenty observations from this curve, compute the mean and standard deviation, and finally compute T. Let's repeat this four thousand times, yielding four thousand values for T, and then plot the results versus what we would get when observations arise from a normal probability curve instead. The results are shown in Figure 5.5. As is evident, the actual probability curve for T differs substantially from the curve we get when sampling from a normal distribution. Consequently, if we compute what we claim is a .95 confidence interval for the mean, the actual probability coverage will be substantially less. In terms of Type I errors, which is just one minus the probability coverage if we test at the .05 level, the actual Type I error probability will be substantially greater than .05. If, for example, we test the hypothesis that the population mean is greater than 1, and we set α equal to .05, the probability of a Type I error will be exactly .05 under normality, but here it is .153 with a sample size of 20 ($n = 20$). For sample sizes of 40, 80, and 160 the actual probability of a Type I error is .149, .124, and .109, respectively. The actual probability of a Type I error is converging to the nominal .05 level as the sample size gets larger, as it should according to the central limit theorem. But we need about two hundred observations to get reasonable control over the probability of a Type I error.

Of course, the seriousness of a Type I error will vary from one situation to the next. However, it is quite common to want the probability of a Type I error to be close to .05. In 1978, J. Bradley argued that in this case, a minimum requirement of any hypothesis testing method is that the actual probability of a Type I error should not be greater than .075 or less than .025. The idea is that if a researcher does not care whether it exceeds .075, then one would want to test at the .075 level, or perhaps at the .1 level, to get more power. (So Bradley assumes that researchers do care; otherwise they would routinely set $\alpha = .075$ rather than .05.) When using Student's T test, there are situations where we need about two hundred observations to ensure that the probability

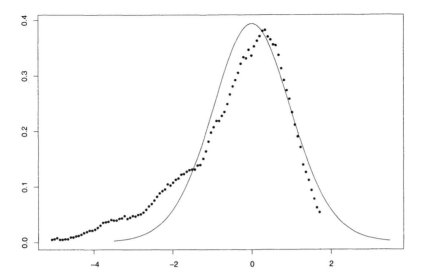

FIGURE 5.5 • The symmetric curve about zero is the probability curve for T when sampling observations from a normal curve. If, however, we sample obervations from the curve in Figure 5.3, the plot of the T values is skewed, as indicated by the dotted line, which is contrary to what is assumed when using conventional hypothesis testing methods for means. The result is poor probability coverage when computing a confidence interval, poor control over the probability of a Type I error, and a biased test when using Student's T.

of a Type I error does not exceed .075. Bradley also argued that ideally, the actual probability of a Type I error should not exceed .055 or be less than .045.

The probability curve in Figure 5.3 has relatively light tails, meaning that the expected number of outliers is small compared to many probability curves we find in applied work. Here is the point. In Chapter 3 we saw that for a skewed, heavy-tailed probability curve, it can take one hundred observations for a plot of the sample means to be well approximated by a normal curve. The pitfall we want to avoid is overgeneralizing and concluding that for skewed, light-tailed probability curves, no practical problems arise because the central limit theorem gives good results even with small sample sizes. As just illustrated, when we must estimate the unknown variance with the sample variance, if we use Student's T, very large sample sizes are required to get accurate results even when outliers tend to be rare.

Here is another result that might appear to be rather surprising and unexpected. Look at the expression for T given by Equation 5.1 and notice that the numerator is $\bar{X} - \mu$. Recall that the expected value of the sample mean is μ. It can be shown that as a result, the expected value of $\bar{X} - \mu$ is zero. That is, $E(\bar{X} - \mu) = 0$. Now the temptation is to conclude that as a consequence, the expected value of T must be zero as well ($E(T) = 0$). Indeed, whenever T is used, this property is being assumed. (It is assumed T is symmetric about zero.) From a technical point of view, the expected value of T must be zero *if* the mean and variance are independent. Under normality this independence is achieved. But for nonnormal probability curves we have dependence, so it does not necessarily follow that the mean of T is zero, an issue that was of concern to Gosset. In fact, there are situations where it is not zero; an example is depicted in Figure 5.5. The mean of T is approximately -0.5.

Why is this important in applied work? Recall that when we test some hypothesis about the mean, we want the probability of rejecting to go up as we move away from the hypothesized value. If we test $H_0 : \mu = 5$, for example, we want to reject if the population mean is actually 7. And if the population mean is 10, we want the probability of rejecting (power) to be even higher. When using T, we do not always get this property, because the expected value of T can differ from zero. Indeed, there are situations where the probability of rejecting is higher when the null hypothesis is true than in situations where it is false. In more practical terms, you have a higher probability of rejecting when nothing is going on than when a difference exists! (In technical terms, there are situations where Student's T is biased.)

One way of trying to salvage Student's T is to suggest, or hope, that the situation just considered never arises in practice. Perhaps we are considering a hypothetical situation that bears no resemblance to any realistic probability curve. But for various reasons, all indications are that the problem is real and more common than might be thought.

Let's elaborate a bit and consider some data from a study on hangover symptoms reported by sons of alcoholics. (The data used here were collected by M. Earleywine.) Using a device called a *bootstrap* (which is described in more detail in Chapter 6), we can estimate the actual distribution of T when it is applied to the data at hand. The estimate is shown in Figure 5.6. As is evident, the estimated probability curve for T differs substantially from the curve we get assuming normality. In other words, if we compute a .95 confidence interval based on T, assuming normality, the expectation is that the actual probability coverage is substantially less than .95. Or, in terms of Type I errors, the actual probability of a Type I error can be substantially higher than .05.

Someone who supports the use of T might argue that this last illustration

Some Practical Problems with Student's T

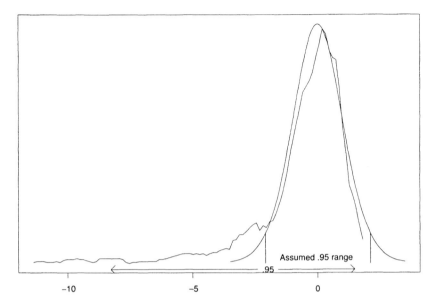

FIGURE 5.6 • When you use Student's T, with twenty observations, you are assuming that the probability curve for T is symmetric about zero, as indicated by the smooth solid line. You are also assuming that with probability .95, the value of T will be between -2.09 and 2.09. This range of values is indicated by the two vertical lines. For the alcohol data, modern methods indicate that, there is a .95 probability that the value of T is between -8.26 and 1.48. Moreover, the probability curve for T is not symmetric, in contrast to the situation where we randomly sample from a normal curve. Also, when using T, you are assuming that its expected value is zero, but here its expected value is approximately -0.71. This results in serious practical problems described in the text.

is not convincing evidence that T is unsatisfactory. Perhaps the bootstrap estimate of the probability curve associated with T is inaccurate and so the illustration is misleading. There is merit to this argument in the sense that the bootstrap estimate is not completely accurate, but despite this, we are not able to salvage T. Based on results in various journal articles, all indications are that the illustration just given *underestimates* the practical problems with Student's T.

Here is yet another property of Student's T that, if not recognized and understood, can lead to practical problems. Consider the values

1 2 3 4 5 6 7 8 9 10 11 12 13 14 15

and imagine we want to test the hypothesis $H_0 : \mu = 5$ with $\alpha = .05$. Using Student's T, the .95 confidence interval for the population mean is (5.52, 10.48); this interval does not contain the value 5, so we reject H_0. Now consider the exact same values, only the last value is increased from 15 to 40. The confidence interval becomes (4.5, 14.8); this interval contains the hypothesized value of 5, so we no longer reject, but the sample mean has increased from 8 to 9.67. That is, the sample mean is further from the hypothesized value, 5. We no longer reject because the sample variance is highly sensitive to outliers, as noted in Chapter 2. For the problem at hand, the outlier inflates the sample variance, and this lengthens the confidence interval for the mean to the point that we no longer reject, even though the sample mean has increased. So in a sense, the sample variance is more sensitive to outliers than the sample mean. If instead the last observation is 100, the sample mean increases to 13.67, yet we still do not reject the hypothesis that $\mu = 5$. Now the .95 confidence interval is (0.25, 27.08). Again, even though outliers inflate the sample mean, we are unable to reject because the outlier has an even bigger impact on the sample variance. Because the sample variance is also going up, we get longer confidence intervals, which now include the hypothesized value, so we are no longer able to reject. So, intuition might suggest that large outliers are good if we want to rule out the possibility that μ is 5, but we see that this is not necessarily the case.

A common suggestion for salvaging Student's T is to transform the data. For example, a typical strategy is to take logarithms of the observations and apply Student's T to the results. Simple transformations sometimes correct serious problems with controlling the probability of a Type I error, but simple transformations can fail to give satisfactory results in terms of Type I errors, particularly in terms of achieving high power and relatively short confidence intervals. There are, however, less obvious transformations that have been found to be relatively effective. We will elaborate on this in Part II.

THE TWO-SAMPLE CASE

In a study conducted by E. Dana, the general goal was to investigate issues related to self-awareness and self-evaluation and to understand the processes involved in reducing the negative effect when people compare themselves to some standard of performance or correctness. A portion of the study hinged on comparing two different (independent) groups of subjects on their ability to keep a portion of an apparatus in contact with a specified target. The amount

THE TWO-SAMPLE CASE

of time the subjects were able to perform this task was measured in hundredths of a second and are reported in Table 5.2.

TABLE 5.2 • SELF-AWARENESS DATA

Group 1:	77 87 88 114 151 210 219 246 253
	262 296 299 306 376 428 515 666 1310 2611
Group 2:	59 106 174 207 219 237 313 365 458 497 515
	529 557 615 625 645 973 1065 3215

How should these two groups be compared? The most common strategy is to test the hypothesis of equal means. If we let μ_1 represent the population mean for the first group and μ_2 the mean for the second, the goal is to test $H_0 : \mu_1 = \mu_2$. If we reject, then conclude that the typical subject in the first group differs from the typical subject in the second. Another approach is to compute a confidence interval for $\mu_1 - \mu_2$. In particular, if the confidence interval does not contain zero, reject H_0 and conclude that $\mu_1 \neq \mu_2$.

A natural estimate of the difference between the population means, $\mu_1 - \mu_2$, is the difference between the corresponding sample means, $\bar{X}_1 - \bar{X}_2$, where $\bar{X}_1$ is the sample mean for the first group and $\bar{X}_2$ is the sample mean for the second. Notice that if we were to repeat an experiment, we would get a different value for this difference. That is, there will be variation among the differences between the sample means if the experiment is repeated many times. It turns out that an expression for this variation can be derived assuming only random sampling; it is

$$\text{VAR}(\bar{X}_1 - \bar{X}_2) = \frac{\sigma_1^2}{n_1} + \frac{\sigma_2^2}{n_2}, \qquad (5.2)$$

where n_1 and n_2 are the corresponding sample sizes. That is, for two independent groups, the variance associated with the difference between the means is the sum of the variances associated with each individual mean. Letting s_1^2 and s_2^2 be the sample variances corresponding to the two groups, this last equation says that we can apply Laplace's general method for computing a confidence interval given by Equation 4.4 of Chapter 4. Here, the $\hat{\theta}$ in Equation 4.4 corresponds to $\bar{X}_1 - \bar{X}_2$, and SE($\hat{\theta}$) is given by the square root of Equation 5.2. Substituting the sample variances for the population variances, Laplace's method for computing a .95 confidence interval suggests using

$$\left(\bar{X}_1 - \bar{X}_2 - 1.96\sqrt{\frac{s_1^2}{n_1} + \frac{s_2^2}{n_2}},\ \bar{X}_1 - \bar{X}_2 + 1.96\sqrt{\frac{s_1^2}{n_1} + \frac{s_2^2}{n_2}} \right)$$

as an approximate .95 confidence interval for the difference between the means. In terms of testing the hypothesis of equal means, reject if $|W| > 1.96$, assuming the probability of a Type I error is to be .95, where

$$W = \frac{\bar{X}_1 - \bar{X}_2}{\sqrt{\frac{s_1^2}{n_1} + \frac{s_2^2}{n_2}}}. \tag{5.3}$$

(Laplace himself used a slight variation of this method for testing the hypothesis of equal means. Here we are applying Laplace's general strategy from a modern perspective.) For the data in Table 5.2, $\bar{X}_1 = 448$, $\bar{X}_1 = 598.6$, $s_1^2 = 353,624.3$, $s_2^2 = 473,804$, and $W = -0.72$. Because $|W|$ is less than 1.96, you would fail to reject the hypothesis that the population means are equal.

If the sample sizes are sufficiently large, then by the central limit theorem, we get a reasonably accurate confidence interval and good control over the probability of a Type I error. But if sample sizes are small, under what conditions can we get exact control over the probability of a Type I error if we use W to test for equal means? A reasonable guess is normality, but this is not enough. If we impose the additional restriction that the population variances be equal, an exact solution can be derived. The resulting method is called the two-sample Student's T test and is based on

$$T = \frac{\bar{X}_1 - \bar{X}_2}{\sqrt{s_p^2 \left(\frac{1}{n_1} + \frac{1}{n_2}\right)}},$$

where

$$s_p^2 = \frac{(n_1 - 1)s_1^2 + (n_2 - 1)s_2^2}{n_1 + n_2 - 2}$$

estimates the assumed common variance. As explained in great detail in every introductory text on applied statistics, the hypothesis of equal means is rejected if $|T| > t$ where t is read from tables based on Student's T distribution. These details are not important here, so for brevity they are not discussed. The main issues, for our current purposes, are the relative merits of using Student's T.

THE GOOD NEWS ABOUT STUDENT'S T

A practical concern is whether Student's T continues to give accurate results when sampling from nonnormal distributions or when the population variances are unequal, contrary to what is assumed. First consider the problem of unequal variances. If observations are sampled from normal distributions and if *equal* sample sizes are used for both groups ($n_1 = n_2$), it can be mathematically verified that the probability of a Type I error will not be too far from the nominal level, no matter how unequal the variances might be. (However, if the common sample size is less than 8, the actual probability of a Type I error can exceed .075 when testing at the .05 level, and some experts have argued that this is unacceptable.) If sampling is from nonnormal distributions that are absolutely identical so that the variances are equal, the probability of a Type I error will not exceed .05 by very much, assuming the method is applied with the desired probability of a Type I error set at $\alpha = .05$. These two results have been known for some time and verified in various studies conducted in more recent years. A casual consideration of the effects of violating assumptions might suggest that Student's T performs well under violations of assumptions. This was certainly the impression generated by research published in the 1950s, and it's a view that dominates applied research today.

THE BAD NEWS ABOUT STUDENT'S T

More recent theoretical and technical advances, and access to high-speed computers have made it possible to study aspects of Student's T that were previously ignored or very difficult to study with the technology of the 1950s. The result is that during the last forty years, several serious problems have been revealed:

- Student's T can have very poor power under arbitrarily small departures from normality relative to other methods one might use. In practical terms, if the goal is to maximize your chances of discovering a true difference between groups, avoid Student's T.

- Probability coverage can differ substantially from the nominal level when computing a confidence interval.

- Power can actually decrease as we move away from the null hypothesis of equal means. (The test is biased.)

- There are general conditions under which Student's T does not even converge to the correct answer as the sample sizes increase.

- Some would argue that Student's T does not control the probability of a Type I error adequately, but others would counter that this is not an issue for reasons elaborated later.

- Population means might poorly reflect the typical subject under study, so the difference between the population means might be misleading in terms of how the typical individual in the first group compares to the typical individual in the second.

The main reason for the first problem and related concerns are described in detail in Chapter 7. In fact, *any* method based on means can result in relatively low power, and among methods based on means, Student's T can be especially bad because of some of its other properties.

To elaborate on the second and fifth issues, consider again the situation where sampling is from normal distributions with unequal variances, but now we consider unequal sample sizes ($n_1 \neq n_2$). Then the actual probability coverage, when attempting to compute a .95 confidence interval, can drop as low as .85. In terms of testing for equal population means, the Type I error probability can exceed .15 when testing at the .05 level. Presumably there are situations where this would be acceptable, but given the very common convention of testing hypotheses at the .05 level, it seems reasonable to conclude that generally this is unsatisfactory in most situations. (Otherwise, everyone would be testing hypotheses with $\alpha = .15$ to get more power.) This suggests using equal sample sizes always, but when we allow the possibility that probability curves are not normal, again poor probability coverage and poor control over the probability of a Type I error can result when the curves differ in shape, even with equal sample sizes.

We saw that for the one-sample T test, power can decrease as we move away from the null hypothesis, for reasons illustrated in Figure 5.5. A similar problem arises here; it is exacerbated in situations where the variances are unequal. That is, if we compare population means that differ, so unknown to us we should reject the hypothesis of equal means, and if simultaneously the variances differ, using Student's T can actually lower power more than various methods that allow the variances to differ. In practical terms, if we want to increase our chances of detecting a true difference between means, it can be to our advantage to abandon T for a so-called heteroscedastic method that allows unequal variances.

A fundamental requirement of any statistical method is that as the sample sizes increase, the inferences we make converge to the correct answer. (In technical terms, we want the method to be asymptotically correct.) In the case of Student's T, if we compute a .95 confidence interval, then as the sample sizes increase, we want the actual probability coverage to converge to .95, and we want the probability of a Type I error to converge to .05. In 1986, N. Cressie and H. Whitford described general conditions under which this property is not achieved by Student's T. This problem can be avoided by abandoning T in favor of W, given by Equation 5.3.

Some would argue that if the probability curves corresponding to two groups have unequal variances, then it is impossible for them to have equal means. This view has been used to argue that if the variances are unequal, then it is impossible to commit a Type I error when testing the hypothesis of equal means because it is impossible for the null hypothesis to be true. The point is that this view takes issue with published journal articles pointing out that the probability of a Type I error is not controlled when using Student's T and variances are unequal. We will not debate the validity of this view here because even if we accept this view, these same journal articles reveal situations where Student's T is biased—power decreases as we move away from the null hypothesis, but eventually it begins to go back up. Surely we want the probability of rejecting to be higher when the population means differ than when they are equal. So regardless of whether one believes that equal means with unequal variances is possible, Student's T has undesirable properties we would like to avoid.

WHAT DOES REJECTING WITH STUDENT'S T TELL US?

Despite the negative features of Student's T, it does tell us one thing when we reject. To explain, let $F_1(x)$ be the probability that an observation randomly sampled from the first group is less than x. For example, if we give a group of subjects some new medication for helping them sleep, $F_1(6)$ is the probability that a randomly sampled subject gets less than six hours of sleep ($x = 6$) and $F_1(7)$ is the probability of less than seven hours. For a control group receiving a placebo, let $F_2(x)$ be the probability of getting less than x hours of sleep. Now imagine we want to test the hypothesis that for any x we choose, $F_1(x) = F_2(x)$. This is a fancy way of saying that the probability curves associated with both groups are identical. When we apply Student's T, all indications are that we can avoid Type I errors substantially larger than .05 when testing this hypothesis. That is, when we reject with Student's T, this is an indication that the probability curves differ in some manner.

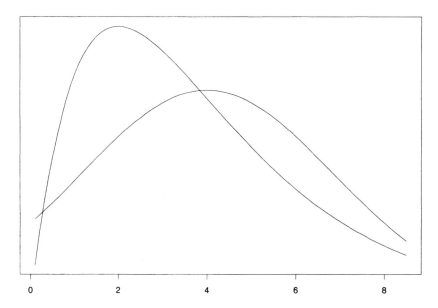

FIGURE 5.7 • An example of two probability curves that obviously differ yet have equal means and variances.

In defense of using Student's T to detect unequal means, some authorities would argue that if the probability curves differ, the means must differ. In theory it is possible to have equal means even when the shapes differ, as depicted in Figure 5.7, but some would argue that the probability of this happening in reality is zero. This argument implies that if we reject with Student's T, surely the means differ. Nevertheless, Student's T is sensitive to many features that characterize the two probability curves, the main reason for rejecting may not be unequal means, but rather unequal variances, differences in skewness, or differences in other measures used to characterize a probability curve. As soon as we conclude that the probability curves differ in some manner, confidence intervals for the difference between the means become suspect. Their probability coverage might be poor. There is doubt about the extent to which the means differ. Is the difference extremely small or substantively important? In some cases Student's T gives accurate results, but the issue is, given some data, can it be trusted to give a good indication of the magnitude of the difference between the population means? Many strategies have been proposed for answering this question. Virtually all of them have been found to be unsatisfactory unless fairly large sample sizes are available, and even then there is some doubt when using Student's T because it does not always converge to the correct answer when the sample sizes increase. The one strategy that seems to have merit is to apply one of the modern methods in Part II of this book. If the results agree with standard techniques, this supports any conclusions one might make about the means. If they differ, there is doubt about whether Student's T can be trusted. Perhaps some other method can be found to empirically validate Student's T, based on data from a given study, but this remains to be seen.

COMPARING MULTIPLE GROUPS

There are standard methods for comparing multiple groups of subjects. They are covered in most introductory texts, but space limitations prevent a thorough description of them here. However, a few words about these standard techniques might be helpful.

Imagine we have three methods for treating migraine: acupuncture, biofeedback, and a placebo. We could compare each pair of groups using, say, Student's T. Of course, for each pair of groups, there is some probability of making a Type I error. A common goal is to ensure that the probability of at least one Type I error is at most .05. So-called multiple comparison procedures have been developed to accomplish this goal. The three best-known are Scheffé's method, Tukey's method, and Fisher's method. All three are based on homoscedasticity and suffer from problems similar to those associated with Student's T, and additional problems not covered here. Generally, as we move from two groups to multiple groups of individuals, it becomes easier to find practical problems when we violate the assumptions of normality and equal variances. There are also methods for testing the hypothesis that all groups have a common mean. (For J groups, the goal is to test $H_0 : \mu_1 = \mu_2 = \cdots = \mu_J$.) The most commonly used technique is called an ANOVA F test and contains Student's T as a special case. Again practical problems are exacerbated when we consider situations where there are more than two groups. The good news is that effective methods for dealing with these problems have been derived.

A SUMMARY OF KEY POINTS

- The basics of hypothesis testing were described. Of particular importance is the fact that when making inferences about μ based on the sample mean, power goes down as σ goes up.

- About one hundred years after Laplace introduced the confidence interval for the mean described in Chapter 4, Gosset attempted to get a more accurate result when sample sizes are small via his Student's T distribution.

- A practical problem with T (given by Equation 5.1) is that its expected value can differ from zero. This is possible because the sample mean and sample variance can be dependent under nonnormality. Consequently, T can provide a biased test. That is, power is not minimized when the null hypothesis is true, meaning that power can actually decrease as we move away from the null hypothesis. Put another way, situations arise where we have a better chance of rejecting when the null hypothesis is false than in situations where the null hypothesis is true. Empirical results were presented, suggesting that in some cases, theoretical problems with T seem to underestimate the extent to which T can be unsatisfactory.

- The two-sample version of Student's T was introduced. Currently, in terms of testing the hypothesis that two probability curves are identical, T seems to control the probability of a Type I error reasonably well. But when the goal is to compare the means, if the probability curves differ, T can be biased and probability coverage (or control over the probability of a Type I error) can be poor. Indeed, there are general conditions where T does not even converge to the correct answer as the sample sizes increase. This last problem can be corrected by replacing T with W given by Equation 5.3, but practical problems remain. (Of particular concern is power under nonnormality for reasons covered in Chapter 7.)

BIBLIOGRAPHIC NOTES

There have been many review papers summarizing practical problems with Student's T and its generalization to multiple groups. The most recent is by Keselman et al. (1998); their paper cites earlier reviews. The result that under general conditions Student's T does not converge to the correct answer was derived by Cressie and Whitford (1986). Taking logarithms of observations is a simple transformation in the sense that each observation is

Bibliographic Notes

transformed in the same manner. (The sample median is not a simple transformation in the sense that some observations are given zero weight, but others are not.) Rasmussen (1989) considered a range of simple transformations with the goal of correcting problems due to nonnormality. He found situations where this approach has value when comparing two groups of individuals, assuming both groups have identical probability curves. But he concluded that this approach does not deal with low power due to outliers. It was stated that Figure 5.6 underestimates problems with Student's T. For more information, the reader is referred to Wilcox (1997). Finally, Glass, Peckham, and Sanders (1972) appears to be one of the earliest papers on the effects of unequal variances when using Student's T (and more generally when using the so-called ANOVA F test). They concluded that violating this assumption is a serious problem, and many more recent studies have shown that having unequal variances is even worse than previously thought. For results on unequal variances when sampling from normal probability curves, see Ramsey (1980).

CHAPTER 6

THE BOOTSTRAP

When testing hypotheses (or computing confidence intervals) with the one-sample Student's T method described in Chapter 5, the central limit theorem tells us that Student's T performs better as the sample size increases. That is, under random sampling the discrepancy between the nominal and actual Type I error probability will go to zero as the sample size goes to infinity. But unfortunately, for reasons outlined in Section 5.3 of Chapter 5, there are realistic situations where about two hundred observations are needed to get satisfactory control over the probability of a Type I error or accurate probability coverage when computing confidence intervals. When comparing the population means of two groups of individuals, using Student's T is known to be unsatisfactory when sample sizes are small or even moderately large. In fact, it might be unsatisfactory no matter how large the sample sizes are because under general conditions it does not converge to the correct answer (Cressie and Whitford, 1986). Switching to the test statistic W given by Equation 5.3, the central limit theorem now applies under general conditions, so using W means we will converge to the correct answer as the sample sizes increase, but in some cases we again need very large sample sizes to get accurate results. (There are simple methods for improving the performance of W using what are called estimated degrees of freedom, but the improvement remains highly unsatisfactory for a wide range of situations.) Consequently, there is interest in finding methods that beat our reliance on the central limit theorem as it applies to these techniques. That is, we would like to find a method that converges to the correct answer more quickly as the sample sizes get large, and such a method is described here.

For various reasons, problems with making accurate inferences about the association between two variables are much more difficult than when comparing measures of location. Equation 4.5 of Chapter 4 described Laplace's method for computing a confidence interval for the slope of the least squares regression. Today a slight variation of this method is used (which was outlined in Section 5.2 of Chapter 5). But even under normality, we will see that the conventional extension of Laplace's method has serious practical problems in terms of achieving accurate probability coverage. A relatively effective method for dealing with this problem is described in this chapter.

In applied work, it is very common to focus attention not on the slope of a regression line, but instead on what is known as Pearson's correlation coefficient. This chapter introduces this coefficient and notes that problems with making inferences about the slope of a regression line extend to it. Fortunately, there are substantially better methods for making inferences about this correlation coefficient, one of which will be described. But unfortunately, there are other more intrinsic problems with this coefficient, described in Chapters 7 and 10, that must also be addressed.

TWO BOOTSTRAP METHODS FOR MEANS

Both theory and simulation studies tell us that a certain form of a relatively modern method generally offers the improvements we seek when computing a confidence interval or testing hypotheses. It is called a *bootstrap method*, two variations of which are covered here. The bootstrap was first proposed by Julian Simon in 1969, and it was discovered independently a short while later by Brad Efron. It was primarily Efron's work that spurred interest in the method. Based on more than a thousand journal articles, all indications are that the bootstrap has great practical value and should be seriously considered in applied work. It is not a panacea, but when combined with other modern insights (covered in Part II), highly accurate results can be obtained in situations where more traditional methods fail miserably.

The basic idea behind all bootstrap methods is to use the data obtained from a study to approximate the sampling distributions used to compute confidence intervals and test hypotheses. When working with means, for example, one version of the bootstrap uses the data to estimate the probability curve associated with T. This is in contrast to the standard strategy of assuming that, due to normality, this probability curve has a specified form that is completely determined by the sample size only. The other version described here, called the *percentile bootstrap*, estimates the sampling distribution of the sample mean instead. Initially, attention is focused on how the bootstrap is used with means, but it generalizes to all of the applied problems considered in this book.

THE PERCENTILE METHOD

To describe the percentile bootstrap method, we begin with a quick review of a sampling distribution as described in Chapter 5. Consider a single population of subjects from which we randomly sample n observations yielding a sample mean, $\bar{X}$. If we obtain a new sample of subjects, in general we get a different sample mean. The sampling distribution of the sample mean refers to the probability that $\bar{X}$ will be less than 2, less than 6, or less than c for any c we might pick. Put another way, there is uncertainty about the value for the sample mean we will get when collecting data, and the sampling distribution of the sample mean refers to the corresponding probability curve.

Next we consider the notion of a sampling distribution from the point of view that probabilities are relative frequencies. If we could repeat a study billions of times, yielding billions of sample means, a certain proportion of the sample means will be less than 2, less than 6, or less than c. If 10 percent of the sample means are less than 2, we say that the probability of getting a sample mean less than 2 is .1. If the proportion less than 6 is 70 percent, we take this to mean that the probability of conducting a study and getting a sample mean less than 6 is .7. What is important from an applied point of view is that if we know these probabilities, we can compute confidence intervals and test hypotheses about the population mean. But obviously we cannot, in most cases, repeat an experiment even twice, let alone billions of times, so it might seem that this description of the sampling distribution has no practical value. However, this description sets the stage for describing the basic strategy behind the bootstrap.

Although we do not know the probability curve that generates observations, it can be estimated from the data at hand. This suggests a method for repeating our experiment without acquiring new observations. For example, imagine we conduct a study aimed at rating the overall mental health of college students, so we administer a standard battery of tests and come up with the following twenty ratings:

2, 4, 6, 6, 7, 11, 13, 13, 14, 15, 19, 23, 24, 27, 28, 28, 28, 30, 31, 43.

The sample mean of these twenty ratings is $\bar{X} = 18.6$. Based on these twenty values, we estimate that the probability of observing the value 2 is 1/20 because exactly one of the twenty observations is equal to 2. In a similar manner, two observations have the value 6, so we estimate that the probability of observing a 6 is 2/20. The probability of getting the value 5 is estimated to be zero because the value 5 was not observed. Obviously, these estimates will differ from the actual probabilities, but the issue is whether these estimated probabilities can be used to get more accurate confidence intervals or better control over the probability of a Type I error.

This estimate of the probability curve suggests the following strategy for estimating the probability curve associated with the sample mean. First, randomly sample, with replacement, twenty observations from the twenty values just listed. In our illustration, this means that each time we sample an observation, the value 2 occurs with probability 1/20, the value 4 occurs with probability 1/20, the value 6 occurs with probability 2/20, and so on. That is, we take the observed relative frequencies to be the probabilities. The resulting twenty observations are called a *bootstrap sample*. For example, we might get

14, 31, 28, 19, 43, 27, 2, 30, 7, 27, 11, 13, 7, 14, 4, 28, 6, 4, 28, 19,

and in fact this bootstrap sample was generated on a computer using the original ratings. The mean of this bootstrap sample, called a *bootstrap sample mean*, is $\bar{X}^* = 18.1$, where the notation $\bar{X}^*$ is used to make a clear distinction with the sample mean from our study, $\bar{X} = 18.6$. If we repeat the process of generating a bootstrap sample, we will get a different bootstrap sample mean. And if we repeat this process say six hundred times we will have six hundred bootstrap sample means. Moreover, if sixty of the six hundred bootstrap sample means are less than 3, then this suggests that if we were to actually repeat our study, as opposed to generating bootstrap samples, our estimate is that with probability 60/600=.1, we will get a sample mean less than 3. Of course, this estimate will be wrong. The only goal for the moment is to convey the flavor of percentile bootstrap: Pretend that the observed values give an accurate estimate of the probability curve and then generate bootstrap sample means in an attempt to approximate the sampling distribution of $\bar{X}$.

Another example might help. Using a computer, let's generate twenty observations from a standard normal curve ($\mu = 0$ and $\sigma = 1$). Theory tells us that the sampling distribution of the sample mean is normal with mean 0 and variance 1/20. But imagine we do not know this and we use the bootstrap to estimate the probability curve using the twenty observations we just generated. This means we repeatedly generate bootstrap samples from these twenty observations and compute a bootstrap sample mean. For illustrative purposes, let's generate six hundred bootstrap sample means. Then we plot the bootstrap sample means and compare the plot to the exact probability curve for the sample mean. That is, we graphically compare the bootstrap estimate of the probability curve to the correct curve. The results are shown in Figure 6.1. As we see, in this particular case the two curves happen to be fairly simi-

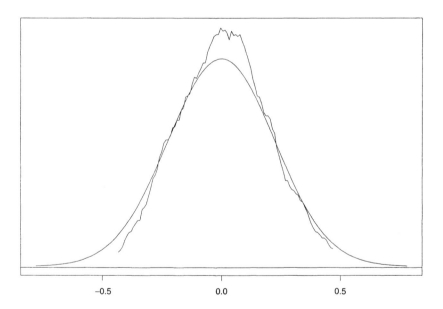

FIGURE 6.1 • Two probability curves. The smooth symmetric curve is what theory tells us we should get for the sampling distribution of the sample mean based on twenty observations. The ragged line is the boostrap approximation of this curve based on twenty observations randomly sampled from a normal curve.

lar. That is, the bootstrap method gives a reasonable approximation of the true probability curve. But of course this one illustration is not convincing evidence that the bootstrap has practical value. Indeed, Figure 6.1 indicates that the plot of the bootstrap sample means does not extend out as far as it should. That is, the probability curve is too light tailed compared to the correct probability curve being estimated. This foreshadows a problem that must be addressed.

Notice that when we generate bootstrap sample means, they will tend to be centered around the sample mean from our study if each bootstrap sample mean is based on a reasonably large number of observations. That is, a version of the central limit theorem applies to the bootstrap sample means. In the last example, the sample mean is $\bar{X} = .01$, so the bootstrap sample means will tend to be centered around .01 rather than the population mean, 0. So, of course, if the sample mean happens to be far from the population mean, the bootstrap sample means will also be centered around a value that is far from the population mean. Despite this, it will generally be the case that the middle 95 percent of the bootstrap sample means will contain the population mean, provided a reasonably large number of observations is available. In our last example, the middle 95 percent of the bootstrap sample means extend from -0.35 to 0.39; this interval contains 0, and this suggests that we should not rule out the possibility that $\mu = 0$.

Suppose we take the middle 95 percent of the bootstrap sample means as a .95 confidence interval for the population mean. In our last example, we are taking the interval $(-0.35, 0.39)$ to be a .95 confidence interval for μ. This is an example of a *percentile bootstrap confidence interval* for the population mean. Furthermore, consider the rule: Reject the hypothesis $H_0 : \mu = 0$ if the bootstrap confidence interval does not contain 0. It can be shown that this rule is reasonable—it can be theoretically justified—provided that the sample size is sufficiently large. That is, if we want the probability of a Type I error to be .05, this will be approximately true if a reasonably large sample size is available.

Returning to the mental health ratings of college students, Figure 6.2 shows a plot of a thousand bootstrap sample means. As indicated, the middle 95 percent of the bootstrap means lie between 13.8 and 23.35. So the interval (13.8, 23.35) corresponds to a .95 percentile bootstrap confidence interval for the unknown population mean.

Unfortunately, when computing confidence intervals for the population mean based on the percentile bootstrap method, large sample sizes are required to get accurate probability coverage, so we have not yet made any

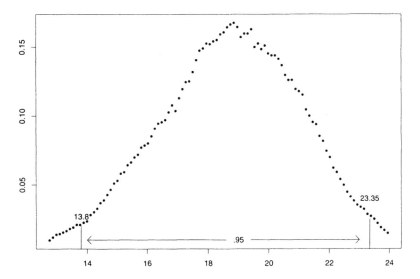

FIGURE 6.2 • A plot of a thousand bootstrap sample means generated from the ratings data. The middle 95 percent of the bootstrap sample means lie between 13.8 and 23.35, suggesting that the interval (13.8, 23.35) be used as an approximate .95 confidence interval for the population mean.

practical progress. Despite this, however, the percentile bootstrap will be seen to have value for other problems we will consider. The only goal here is to describe the percentile bootstrap for the simplest case.

THE PERCENTILE t BOOTSTRAP

Another form of the bootstrap method arises as follows. Recall that when computing a confidence interval for μ, a solution is obtained by assuming that

$$T = \frac{\bar{X} - \mu}{s/\sqrt{n}}$$

has a Student's T distribution. If, for example, $n = 25$, it can be shown that when sampling from a normal curve, there is a .95 probability that T will be between -2.064 and 2.064. This result can be used to show that a .95 confidence interval for the population mean is

$$\left(\bar{X} - 2.064\frac{s}{\sqrt{n}}, \bar{X} + 2.064\frac{s}{\sqrt{n}}\right)$$

when sampling from a normal distribution. The point is that assuming normality provides an approximation of the probability curve for T, which in turn yields an approximate .95 confidence interval when sampling from nonnormal distributions. But as previously indicated, a practical concern is that this approximation of the probability curve for T performs poorly in some cases, which in turn means we get inaccurate confidence intervals, poor control over the probability of a Type I error, and undesirable power properties. If we could determine the probability curve for T without assuming normality, the problems associated with Type I errors and probability coverage would be resolved. What we need is a better way of approximating the distribution of T.

A percentile t bootstrap approximates the distribution of T as follows. First, obtain a bootstrap sample as was done when we applied the percentile bootstrap method. For this bootstrap sample, compute the sample mean and standard deviation and label the results $\bar{X}^*$ and s^*. As an illustration, consider again the study aimed at assessing the overall mental health of college students based on the twenty ratings

2, 4, 6, 6, 7, 11, 13, 13, 14, 15, 19, 23, 24, 27, 28, 28, 28, 30, 31, 43.

For the bootstrap sample previously considered, namely,

14, 31, 28, 19, 43, 27, 2, 30, 7, 27, 11, 13, 7, 14, 4, 28, 6, 4, 28, 19,

we get $\bar{X}^* = 18.1$, and a bootstrap standard deviation of $s^* = 11.57$. Next, compute

$$T^* = \frac{\bar{X}^* - \bar{X}}{s^*/\sqrt{n}}. \tag{6.1}$$

In the illustration,

$$T^* = \frac{18.6 - 18.1}{11.57/\sqrt{20}} = 0.19.$$

Repeat this process B times, each time computing T^*. Figure 6.3 shows a plot of $B = 1,000$ values obtained in this manner. These B values provide an approximation of the distribution of T without assuming normality.

As indicated by Figure 6.3, 95 percent of these one thousand values lie between -2.01 and 2.14. If instead we assume normality, then 95 percent of the T values would be between -2.09 and 2.09. So in this particular case, there is little difference between the percentile t bootstrap and assuming normality.

Here is a summary of how to compute a .95 confidence interval for the mean using the percentile t bootstrap method:

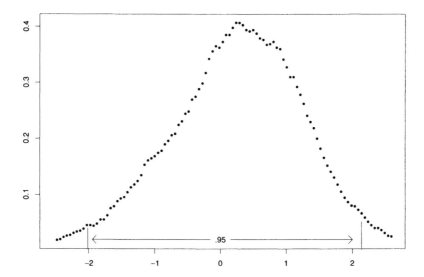

FIGURE 6.3 • A bootstrap estimate of the sampling distribution of T based on the ratings data. The middle 95 percent of the bootstrap T values lie between -2.01 and 2.14. When sampling from a normal distribution, the T values will lie between -2.09 and 2.09 with probability $.95$. In this case, the percentile t bootstrap is in close agreement with what we get when assuming normality.

1. Compute the sample mean, $\bar{X}$, and standard deviation, s.

2. Generate a bootstrap sample by randomly sampling with replacement n observations from $X_1, \ldots, X_n$, yielding $X_1^*, \ldots, X_n^*$.

3. Use the bootstrap sample to compute T^* given by Equation 6.1.

4. Repeat steps 2 and 3 B times yielding $T_1^*, \ldots, T_B^*$. For $\alpha = .05$, B must be fairly large when working with means. Based on results in Hall (1986), the choice $B = 999$ is recommended rather than the seemingly more natural choice of $B = 1,000$. For n small (less than 100), unsatisfactory probability coverage can result when working with means, and increasing B seems to offer little or no advantage. But smaller values for B give good results for some of the methods described in subsequent chapters.

5. Write the bootstrap T^* values in ascending order, yielding $T_{(1)}^* \leq \cdots \leq T_{(B)}^*$.

6. Set $L = .025B$ and $U = .975B$ and round both L and U to the nearest integer.

The bootstrap percentile t confidence interval for μ (also called the bootstrap-t interval) is

$$\left(\bar{X} - T^*_{(U)}\frac{s}{\sqrt{n}},\ \bar{X} - T^*_{(L)}\frac{s}{\sqrt{n}}\right).$$

(For readers familiar with basic statistics, $T^*_{(L)}$ will be negative, and that is why $T^*_{(L)}s/\sqrt{n}$ is subtracted from the sample mean. Also, it might seem that $T^*_{(L)}$ should be used to define the lower end of the confidence interval, but it can be seen that this is not the case.) In the illustration where $\bar{X} = 18.6$ and $s = 11.4$, a .95 confidence interval for the mean based on the percentile t method (using software mentioned in the final chapter) is

$$\left(18.6 - 2.08\frac{11.4}{\sqrt{20}},\ 18.6 + 2.55\frac{11.4}{\sqrt{20}}\right) = (10.2, 26.9).$$

If instead normality is assumed, the confidence interval is (13.3, 23.9).

An important issue is whether the percentile t bootstrap ever gives a substantially different result than assuming normality. If it never makes a difference, of course, there is no point in abandoning Student's T for the percentile t bootstrap. The following example, based on data taken from an actual study, illustrates that substantial differences do indeed occur.

M. Earleywine conducted a study on hangover symptoms after consuming a specific amount of alcohol in a laboratory setting. For one of the groups, the results were

0, 32, 9, 0, 2, 0, 41, 0, 0, 0, 6, 18, 3, 3, 0, 11, 11, 2, 0, 11.

(These data differ from the data used to create Figure 5.6, but they are from the same study.) Figure 6.4 shows the bootstrap distribution of T based on $B = 999$ bootstrap samples. The middle 95 percent of the T^* values are between -4.59 and 1.61. If we assume normality, then by implication the middle 95 percent of the T values will be between -2.09 and 2.09 instead. Figure 6.4 also shows the distribution of T assuming normality. As is evident, there is a substantial difference between the two methods. The .95 confidence interval based on the percentile t method is $(-3.13, 11.5)$, and it is $(2.2, 12.8)$ when assuming normality.

Using yet another set of data from the same study, namely,

0, 0, 0, 0, 0, 0, 0, 0, 1, 8, 0, 3, 0, 0, 32, 12, 2, 0, 0, 0,

the middle 95 percent of the T values are estimated to lie between -13.6 and 1.42. That is, there is an even bigger discrepancy between the bootstrap and what we get assuming normality.

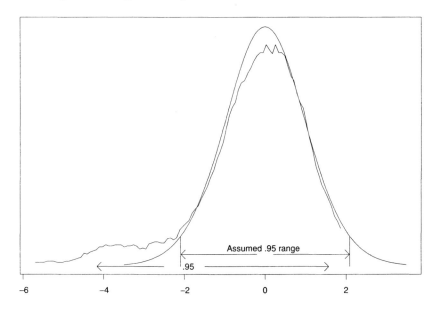

FIGURE 6.4 • In applied work, the approximation of the probability curve for T, based on the bootstrap, can differ substantially from the approximation based on the normal curve.

Because the two methods for computing confidence intervals can differ substantially, there is the issue of which you should use. If distributions are normal, Student's T offers a very slight advantage. In general, however, including situations where distributions are nonnormal, it seems to never offer a substantial advantage. In contrast, the percentile t bootstrap offers a substantial advantage over Student's T in various realistic situations, so it deserves serious consideration in applied work. A reasonable suggestion is to use Student's T if a distribution seems to be approximately normal, or if the sample size is sufficiently large, but this is too vague. How close to a normal distribution must it be? Currently, there is no satisfactory answer to this question. We can make an educated guess that with two hundred observations, Student's T will perform as well as the bootstrap in most situations, but there is no proof that this is the case. The best advice seems to be to always use the percentile t bootstrap when making inferences about a mean.

TESTING HYPOTHESES

Bootstrap confidence intervals can be used to test hypotheses. Basically, you proceed as indicated in Chapter 5 and reject if the hypothesized value is not contained in the confidence interval. If in the last example there is interest in testing $H_0 : \mu = 12$, the percentile t bootstrap method would reject with $\alpha = .05$ because the .95 confidence interval $(-3.13, 11.5)$ does not contain the hypothesized value, 12. In contrast, Student's T would not reject because its .95 confidence interval (2.2, 12.8) contains 12.

WHY DOES THE PERCENTILE t BOOTSTRAP BEAT STUDENT'S T?

The increased accuracy of the percentile t bootstrap versus Student's T is not surprising based on certain theoretical results. To explain, first consider Student's T. Whenever we use it to compute a .95 confidence interval, there is generally some discrepancy between .95 and the actual probability coverage. Or, when testing some hypothesis with the goal that the probability of a Type I error be .05, the actual probability of a Type I error generally differs from .05 due to nonnormality. Mathematicians are able to characterize how quickly this discrepancy goes to zero as the sample size increases, the rate is $1/\sqrt{n}$. That is, $1/\sqrt{n}$ goes to zero as the sample size, n, gets large, and this provides some indication of how quickly errors made with Student's T go to zero. This does *not* mean that the difference between the nominal and actual Type I error probabilities is $1/\sqrt{n}$—we have already seen that in some cases we need two hundred observations when using Student's T. But this ratio is used by mathematicians to measure how well a given method performs.

The point is that when using the percentile t bootstrap, the discrepancy between the actual and nominal Type I error probabilities goes to zero at the rate $1/n$—it goes to zero more quickly than when using Student's T. So from a theoretical perspective, the percentile t beats Student's T. Unfortunately, this by itself is not convincing evidence that in applied work, the percentile t beats Student's T when sampling from nonnormal probability curves because with small sample sizes, it is not remotely obvious how the performance of the percentile t will compare to Student's T based on this theoretical result. The theory is based on a large-sample comparison of the methods and might give a very poor indication of how they compare when sample sizes are small or even moderately large. Moreover, this result does *not* tell us how large a sample we need to get accurate results with the percentile t bootstrap. Quantitative experts use simulation studies to answer these questions. The good news is that in simulation studies, typically the percentile t bootstrap performs about as well as, and in some cases much better than, Student's T. Moreover, there are no indications that Student's T ever offers a substantial improvement over the percentile t bootstrap. The bad news is that when working with the mean, although we get increased accuracy, situations arise where the control over the probability of a Type I error remains unsatisfactory. For example, in Section 5.3 of Chapter 5 we saw a situation where, when testing a hypothesis about the mean, we needed about two hundred observations to get accurate control over the probability of a Type I error. If we switch to the percentile t bootstrap, we reduce the required number of observations to one hundred. So substantial progress has been made, but more needs to be done. We have seen that in some situations, Student's T is biased; its power might actually decline as we move away from the null hypothesis. The percentile t bootstrap reduces this problem as well, but unfortunately it does not eliminate it. Moreover, when using the percentile t bootstrap with means, a fundamental problem described in Chapter 7 remains.

COMPARING TWO INDEPENDENT GROUPS

The bootstrap methods described in the previous section are easily extended to the problem of comparing two independent groups. Recall from Chapter 5 that Student's T for comparing means assumes groups have equal variances, even when the means differ. One possibility is to use a bootstrap analog of Student's T test, but this approach is not described because it does not correct the technical problems associated with violating the assumption of equal variances. One of the better methods for comparing means is to use a percentile t bootstrap based on the test statistic W given by Equation 5.3. To compute a .95 confidence interval for $\mu_1 - \mu_2$, proceed as follows:

1. Compute the sample mean and standard deviation for each group and label the results $\bar{X}_1$ and s_1 for group 1, and $\bar{X}_2$ and s_2 for group 2. Set $d_1 = s_1^2/n_1$ and $d_2 = s_2^2/n_2$, where n_1 and n_2 are the sample sizes.

2. Generate a bootstrap sample from the first group, compute the bootstrap sample mean and standard deviation, and label the results $\bar{X}_1^*$ and s_1^*. Do the same for the second group yielding $\bar{X}_2^*$ and s_2^*. Set $d_1^* = (s_1^*)^2/n_1$ and $d_2^* = (s_2^*)^2/n_2$.

3. Compute
$$W^* = \frac{(\bar{X}_1^* - \bar{X}_2^*) - (\bar{X}_1 - \bar{X}_2)}{\sqrt{d_1^* + d_2^*}}.$$

4. Repeat steps 2 and 3 B times yielding $W_1^*, \ldots, W_B^*$. For a Type I error of .05, which corresponds to computing a .95 confidence interval, $B = 999$ is recommended. (Smaller values for B can be used in situations to be covered.)

5. Put the $W_1^*, \ldots, W_B^*$ values in ascending order, yielding $W_{(1)}^* \leq \cdots \leq W_{(B)}^*$.

6. Set $L = .025B$, $U = .975B$ and round both L and U to the nearest integer.

The bootstrap percentile t confidence interval for $\mu_1 - \mu_2$ is

$$\left(\bar{X}_1 - \bar{X}_2 - W_{(U)}^* \sqrt{d_1 + d_2},\ \bar{X}_1 - \bar{X}_2 - W_{(L)}^* \sqrt{d_1 + d_2} \right).$$

HYPOTHESIS TESTING

Reject $H_0 : \mu_1 = \mu_2$, the hypothesis that two groups have equal means, if the confidence interval just computed does not contain 0. If, for example, the confidence interval is (1.2, 2.3), the estimate is that the difference between the means ($\mu_1 - \mu_2$) is at least 1.2, so the situation $\mu_1 - \mu_2 = 0$ seems unlikely in light of the data.

It is stressed that if groups do not differ and if they have identical probability curves, bootstrap methods offer little or no advantage over nonbootstrap methods in terms of Type I errors. However, this does not salvage nonbootstrap methods because, of course, you do not know whether the groups differ. If the groups do differ, the bootstrap tends to provide more accurate confidence intervals. In some situations the improvement is substantial. As in the previous section, it seems that standard methods offer a minor advantage in some cases but never a major one. Consequently, the percentile t bootstrap is recommended for comparing means.

REGRESSION

Equation 4.5 of Chapter 4 described Laplace's method for computing a confidence interval for the slope of a regression line based on the least squares estimator. Today a slight variation of this method is routinely used and is described in most introductory texts. The method is again based on the least squares estimate of the slope, but the value 1.96 in Equation 4.5 is replaced by a larger value, the magnitude of which depends on the sample size and is read from tables of Student's T distribution (with $n - 2$ degrees of freedom). Often this method is used to test $H_0 : \beta_1 = 0$, the hypothesis that the slope of the regression line is zero, and this hypothesis is rejected if the confidence interval for the slope does not contain zero. Unfortunately, this relatively simple method can be disastrous in terms of Type I errors and probability coverage, even under normality. If, for example, there is heteroscedasticity (meaning that the variance of the outcome measure, Y, changes with the value of the predictor, X, as described in Chapter 4), the actual probability of a Type I error can exceed .5 when testing at the .05 level. Some authorities would counter that in applied work, it is impossible to simultaneously have heteroscedasticity and a slope of zero. That is, Type I errors are never made when there is heteroscedasticity because the null hypothesis of a zero slope is virtually impossible. Even if we accept this argument, another concern is that heteroscedasticity can mask an association of practical importance. Serious problems arise even under normality because Laplace's method and its modern extension assume homoscedasticity, which leads to an expression for the variance of the least squares estimate of the slope. The concern is that under heteroscedasticity this expression is no longer valid, and this leads to practical problems that were impossible to address in an effective manner until fairly recently.

The bootstrap can be extended to the problem of computing a confidence interval for the slope of a regression line in a manner that takes heteroscedasticity into account. To apply it to the observed n pairs of observations available to us, we begin by randomly sampling, with replacement, n pairs of observations from the data at hand. To illustrate the process, again consider Boscovich's data on meridian arcs, which were described in Chapter 2. For convenience, we list the five observed points here: (.0000, 56,751), (.2987, 57,037), (.4648, 56,979), (.5762, 57,074), and (.8386, 57,422). A bootstrap sample consists of randomly selecting, with replacement, five pairs of observations from the five pairs available to us. Using a computer, the first pair we select might be (.4648, 56,979). When we draw the second pair of values, with probability 1/5 we will again get the pair (.4648, 56,979). More generally, when we have n pairs of observations, a bootstrap sample consists of randomly selecting a pair of points, meaning each point has probability $1/n$ of being chosen, and repeating this process n times.

For completeness, there is another approach to generating bootstrap samples based on residuals. Theoretical results tell us, however, that it should not be used when there is heteroscedasticity, and studies that assess how the method performs with small sample sizes also indicate that the method can be highly unsatisfactory. We could test the assumption of homoscedasticity, but it is unknown how to determine whether such tests have enough power to detect situations where this assumption should be discarded. Consequently, details about this other bootstrap method are not given here.

Once we have a bootstrap sample of n pairs of points, we can compute a bootstrap estimate of the slope. For example, if the bootstrap sample for Boscovich's data happens to be (.4648, 56,979), (.0000, 56,751), (.8386, 57,422), (.4648, 56,979), and (.0000, 56,751), the least squares estimate of the slope based on these five pairs of observations is 737.4. That is, 737.4 is a bootstrap estimate of the slope. If we obtain a new bootstrap sample, typically it will differ from the first bootstrap sample and yield a new bootstrap estimate of the slope.

Next we proceed as with the percentile bootstrap method for the mean. That is, we generate many bootstrap estimates of the slope and take the middle 95 percent to be a .95 confidence interval for the true slope. This method improves on the conventional approach based on Student's T, but unfortunately it requires about two hundred fifty pairs of observations to get reasonably accurate results over a wide range of situations. There is, however, a simple modification of the method that has been found to perform well when sample sizes are small. It is based on the observation that for a given sample size, the actual probability coverage obtained with the percentile bootstrap method is fairly stable. If, for example, the actual probability coverage is .9 under normality, it will be approximately .9 when sampling from a nonnormal curve instead. This suggests that if we expand our confidence interval so that under normality the actual probability coverage will be .95, then it will be about .95 under nonnormality, and this has been found to be true for a wide range of situations. In terms of testing hypotheses, the actual probability of a Type I error will be reasonably close to .05.

The method is implemented as follows. First generate five hundred ninety-nine bootstrap estimates of the slope and label them $\hat{\beta}_1^*, \hat{\beta}_2^*, \ldots, \hat{\beta}_{599}^*$. Next, put these values in order and label them $\hat{\beta}_{(1)}^* \leq \hat{\beta}_{(2)}^* \leq \cdots \leq \hat{\beta}_{(599)}^*$. The

.95 confidence interval for slope, based on the least squares estimator, is $(\hat{\beta}^*_{(a)}, \hat{\beta}^*_{(c)})$, where for $n < 40$, $a = 7$ and $c = 593$; for $40 \leq n < 80$, $a = 8$ and $c = 592$; for $80 \leq n < 180$, $a = 11$ and $c = 588$; for $180 \leq n < 250$, $a = 14$ and $c = 585$; and for $n \geq 250$, $a = 15$ and $c = 584$. If, for example, $n = 20$, the lower end of the .95 confidence interval is given by $\hat{\beta}^*_{(7)}$, the seventh of the five hundred ninety-nine bootstrap estimates after they are put in ascending order. This method becomes the standard percentile bootstrap procedure when $n \geq 250$. It is stressed that although this method performs fairly well in terms of Type I errors, any method based on the least squares estimator might be unsatisfactory for reasons outlined in Section 7.7 of Chapter 7.

The success of the method just described, in terms of Type I errors, is somewhat surprising. As noted in Chapter 2, the least squares estimate of the slope is just a weighted mean of the outcome (Y) values. This suggests that the modified percentile bootstrap method for the slope might also work well when trying to test hypotheses about the population mean using $\bar{X}$. But it has been found that this is not the case. Using the percentile bootstrap to compute a confidence interval for μ is very unstable, so any simple modification along the lines considered here is doomed to failure.

To illustrate the practical difference between the conventional method for computing a confidence interval for the slope, and the percentile bootstrap, again consider Boscovich's data. The conventional method yields a .95 confidence interval of (226.58, 1220.30). In contrast, the modified percentile bootstrap method gives (-349.19, 1237.93). The upper ends of the two confidence intervals are similar, but the lower ends differ substantially, so we see that the choice of method can make a practical difference.

CORRELATION AND TESTS OF INDEPENDENCE

When studying the association between two variables, it is common for applied researchers to focus on what is called *Pearson's correlation coefficient* rather than on the least squares estimate of the slope. The two methods are intimately connected, but the information conveyed by the correlation coefficient differs from the least squares estimate of the slope (except in a special case that will be discussed in Chapter 10). The immediate goal is to introduce this measure of association and note that again heteroscedasticity plays a role when applying a conventional method aimed at establishing whether two variables are dependent.

Given n pairs of observations, $(X_1, Y_1), \ldots, (X_n, Y_n)$, the sample covariance between X and Y is

$$\text{COV}(X, Y) = \frac{1}{n-1}[(X_1 - \bar{X})(Y_1 - \bar{Y}) + \cdots + (X_n - \bar{X})(Y_n - \bar{Y})].$$

For Boscovich's data, $\bar{X} = 0.436$, $\bar{Y} = 57052.6$, there are $n = 5$ pairs of points, so the sample covariance is

$$\frac{1}{5-1}[(0.000 - 0.436)(56751 - 57052.6)$$
$$+ \cdots + (0.8386 - 0.436)(57422 - 57052.6)] = 70.8.$$

Covariance is a generalization of the sample variance in the sense that the covariance of the variable X with itself, $\text{COV}(X,X)$, is just s_x^2, the sample variance of the X values.

The estimate of Pearson's correlation coefficient is

$$r = \frac{\text{COV}(X, Y)}{s_x s_y}, \tag{6.2}$$

where s_x and s_y are the sample standard deviations corresponding to X and Y, respectively. For Boscovich's data, $r = 0.94$. The population analog of r (the value of r if all subjects or objects could be measured) is typically labeled ρ. It can be shown that the value of ρ always lies between -1 and 1 and that when X and Y are independent, $\rho = 0$. So if one can reject the hypothesis that $\rho = 0$, dependence between X and Y is implied. In addition, it can be shown that if $\rho > 0$ the least squares regression line will have a positive slope (meaning that according to the least squares line, Y tends to increase as X increases), and if $\rho < 0$, the reverse is true.

The conventional test of $H_0 : \rho = 0$ is derived under the assumption that X and Y are independent. An implication of this assumption is that the error term, when predicting Y from X, is homoscedastic. This implication makes it possible to derive a commonly used test statistic:

$$T = r\sqrt{\frac{n-2}{1-r^2}}. \tag{6.3}$$

Under normality, and when $\rho = 0$, T has a Student's T distribution (with $n - 2$ degrees of freedom). If, for example, twenty-six pairs of observations are used to compute r, then with probability .95, T will have a value between -2.064 and 2.064. So if we reject when $|T| > 2.064$, the probability of a

Type I error will be .05, still assuming normality. (Computational details can be found in virtually any introductory text.)

There are many pitfalls associated with r and ρ, most of which are described in Part II of this book. For now we focus on an issue related to testing the hypothesis that $\rho = 0$. If indeed the population correlation coefficient is zero, does this imply independence? The answer is no, not necessarily. For example, suppose X and Y are independent, standard normal random variables when $X \leq 1$, but for $X > 1$, the standard deviation of Y is X. Given that $X = .5$, say, the standard deviation of Y is 1, but if $X = 2$, the standard deviation of Y is 2. Then $\rho = 0$, yet X and Y are dependent because knowing the value of X can alter the probabilities associated with Y.

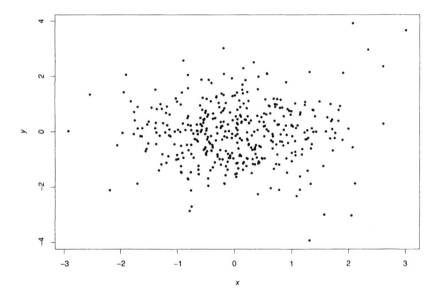

FIGURE 6.5 • This cloud of points was generated from a situation where X and Y are dependent but $\rho = 0$. The cloud of points might suggest that there is independence, but for $X > 1$, the variance of Y increases with X.

Figure 6.5 shows a plot of four hundred values generated on a computer in the manner just described. That is, four hundred pairs of values for both X and Y were generated from a standard normal probability curve, and if the value for X is greater than one, Y was multiplied by X. Notice that there is little or no indication that there is an association between X and Y. Not only is the population correlation zero, but the population slope of the least squares regression line (β_1) is also zero. Yet, when using T to test the hypothesis of a zero correlation at the .05 level, the actual probability of rejecting is .15. In this particular case, when we reject, the correct conclusion is that X and Y are dependent, but it would be incorrect to conclude that $\rho \neq 0$ and that the mean of Y increases or decreases with X. The problem is that the test of $H_0 : \rho = 0$ is based on the assumption that there is homoscedasticity, but in reality there is heteroscedasticity, and this causes the probability of rejecting to exceed .05 even though the hypothesis being tested is true.

To take a more extreme example, suppose e and X are independent, standard normal variables and that $Y = |X|e$. (So to generate a value for Y, we generate a value from the standard normal probability curve, call it e, generate a value for X independent of e, and then set $Y = |X|e$.) Then again X and Y have zero correlation even though X and Y are dependent. (They are dependent because the variance of Y changes with X.) Now when testing $H_0 : \rho = 0$ at the .05 level, the actual probability of a Type I error, based on a sample of $n = 20$ points, is .24. Increasing the sample size to 400, the probability of rejecting is .39. That is, the probability of rejecting *increased* even though the hypothesis of a zero correlation is true. That is, we are more likely to reject with a larger sample size even though the hypothesis about Pearson's correlation is true and should not be rejected. The probability of incorrectly rejecting increases due to heteroscedasticity. So when we reject, it is correct to conclude that X and Y are dependent, but we must be careful about any inferences we draw about how X and Y are related. If we reject and $r > 0$, for example, it is certainly true that the estimated least squares regression will have a positive slope, but this does not necessarily mean that it is generally the case that the expected (or average) value of Y increases with X. In our example it does not, and we will see a variety of other ways the value of r might mislead us.

Put another way, when we reject the hypothesis that $\rho = 0$, this might be because $\rho \neq 0$, but an additional factor causing us to reject might be heteroscedasticity. It turns out that we can separate the influence of these two factors using the (modified) percentile bootstrap method used to compute a confidence interval for the slope of the least squares regression line. The method is exactly the same as before, but for each bootstrap sample we simply compute the correlation coefficient rather than the least squares estimate of the slope. In particular, generate a bootstrap sample by sampling with replacement n pairs of observations. Then compute Pearson's correlation coefficient and label it r^*. Repeat this five hundred ninety-nine times and label the resulting correlation coefficients $r_1^*, \ldots, r_{599}^*$. Next, put these values in order and label them $r^*_{(1)} \le \cdots \le r^*_{(599)}$. The .95 confidence interval for ρ is $(r^*_{(a)}, r^*_{(c)})$, where for $n < 40$, $a = 7$ and $c = 593$; for $40 \le n < 80$, $a = 8$ and $c = 592$; for $80 \le n < 180$, $a = 11$ and $c = 588$; for $180 \le n < 250$, $a = 14$ and $c = 585$; and for $n \ge 250$, $a = 15$ and $c = 584$.

In our illustrations where T, given by Equation 6.3, does not control the probability of a Type I error, the probability of rejecting $H_0 : \rho = 0$ is close to .05, as intended, even though there is heteroscedasticity. That is, the bootstrap separates inferences about the population correlation coefficient from a factor that contributes to our probability of rejecting. Insofar as we want a test of $H_0 : \rho = 0$ to be sensitive to ρ only, the bootstrap is preferable.

Some might argue that it is impossible to have heteroscedasticity with ρ exactly equal to zero. That is, we are worrying about a theoretical situation that will never arise in practice. Regardless of whether one agrees with this view, the sensitivity of the T test of $H_0 : \rho = 0$ is relatively minor compared to other problems to be described. What might be more important is whether heteroscedasticity masks an association. That is, from an applied point of view, perhaps there are situations where we fail to reject with T, not because there is no association but because there is heteroscedasticity. But even if this concern has no relevance, several other concerns have been found to be extremely important in applied work.

We mention one of these concerns here, but we must postpone a discussion of the others until more basic principles are described. (These additional concerns are described in Chapters 7 and 10.) The concern illustrated here is that the sample breakdown point of the correlation coefficient, r, is only $1/n$. That is, one point, properly placed, can cause the correlation coefficient to take on virtually any value between -1 and 1, so care must be taken when interpreting that value of r.

Figure 6.6 shows a scatterplot of the logarithm of the surface temperature of forty-seven stars versus the logarithm of their light intensity. The scatterplot suggests that, generally, there is a positive association between temperature and light intensity, yet $r = -.21$. The value of r is negative because of the four outliers in the upper-left corner of Figure 6.6. That is, from a strictly numerical point of view, r suggests there is a negative association, but for the bulk of the points the scatterplot indicates that the reverse is true. One could argue that these outliers are interesting because they hint at the possibility that the association changes for relatively low X values. It cannot be emphasized too strongly that, of course, outliers can be interesting. Theoreticians who work on robust statistical methods assume that this is obvious. In this particular case, the outliers happen to be red giant stars, and perhaps this needs to be taken into account when studying the association between light intensity and surface temperature. Note that for the six X values less than 4.2 (shown in the left portion of Figure 6.6), the scatterplot suggests that there might be a negative association; otherwise the association appears to be positive. So perhaps there is a nonlinear association between temperature and light intensity. That is, fitting a straight line to these data might be ill advised when considering the entire range of X values available to us.

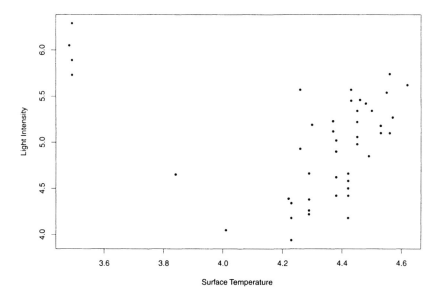

FIGURE 6.6 • A scatterplot of the star data illustrating that outliers can greatly influence r. Here, r is negative even though for most of the points there is a positive association.

We have just seen that simply looking at a scatterplot can give us an indication that the value of r is misleading. But in Chapter 7 we will see examples where even a scatterplot is potentially deceptive. (See, in particular, the discussion of Figures 7.10 and 7.11.) Understanding the association between two or more variables requires a library of tools. Subsequent chapters will elaborate on this important point and illustrate some of the tools one might use.

A SUMMARY OF KEY POINTS

- Both theory and simulations indicate that the percentile t bootstrap beats our reliance on the central limit theorem when computing confidence intervals for the population mean. Practical problems with Student's T are reduced but not eliminated. It was illustrated that Student's T and the percentile t bootstrap can yield substantially different results.

- The percentile bootstrap method is not recommended when working with the sample mean, but it has practical value when making inferences about the slope of a regression line. For example, when using the least squares estimate of the slope, the modified percentile bootstrap method in Section 6.5 provides relatively accurate probability coverage even under heteroscedasticity. Even under normality, heteroscedasticity can invalidate the conventional method based on Student's T.

- Heteroscedasticity also affects the conventional T test of the hypothesis that $\rho = 0$. Again this problem is corrected substantially by using the modified percentile bootstrap method in Section 6.6.

BIBLIOGRAPHIC NOTES

There are many variations of the bootstrap beyond those described here. For books completely dedicated to bootstrap methods, see Efron and Tibshirani (1993), Davison and Hinkley (1997), and Shao and Tu (1995). In this chapter a method was described for obtaining bootstrap samples in regression when there is heteroscedasticity. For theoretical results motivating this approach, see Wu (1986). The modification of the percentile method for computing a confidence for the slope of a least squares regression line comes from Wilcox (1996a).

CHAPTER 7

A FUNDAMENTAL PROBLEM

In the year 1960, John Tukey published a paper on the so-called *contaminated* or *mixed normal distribution* that would have devastating implications for conventional inferential methods based on means. Indeed, any method based on means would, by necessity, suffer from a collection of similar problems. Tukey's paper had no immediate impact on applied work, however, because it was unclear how to deal with the practical implications of his paper. It served as the catalyst for the theory of robustness that was subsequently developed by Frank Hampel and Peter Huber, and their results laid the foundation for getting practical solutions to the problems revealed by Tukey.

The fundamental problem revealed by Tukey's paper is that arbitrarily small departures from normality can have a large impact on the population variance. This has important implications about any method based on means. To elaborate, we begin by describing the mixed normal in concrete terms. Imagine we have a measure of self-esteem and consider two populations of subjects: schizophrenics and nonschizophrenics. For illustrative purposes, suppose that our measure of self-esteem for the nonschizophrenics has a standard normal distribution, in which case the population mean (μ) is 0 and the standard deviation (σ) is 1. For the schizophrenics, again suppose the distribution of self-esteem scores is normal with mean $\mu = 0$, but the standard deviation is $\sigma = 10$. Further suppose that 10 percent of the population is schizophrenic and we mix the two groups together. That is, when we sample an adult, there is a 10 percent chance of getting a schizophrenic, so there is a 90 percent chance that we will sample a nonschizophrenic, meaning that there is a 90 percent chance that we are sampling from a standard normal distribution. Similarly, there is a 10 percent chance that we sampled from a normal distribution with mean $\mu = 0$ and standard deviation $\sigma = 10$.

Now, when we sample an adult, there is a certain probability that the self-esteem score will be less than any constant c we might choose. If we are told that the adult is not schizophrenic, we can determine the probability of a self-esteem score being less than c using basic methods for determining probabilities associated with normal probability curves. But suppose we want to determine the probability without knowing if the adult is schizophrenic. In symbols, how do we determine $P(X \leq c)$ (the probability that an observation will be less than or equal to c) if we do not know whether we are sampling from a standard normal distribution or a normal distribution with $\sigma = 10$? There is a method for answering the question exactly (see e.g., Rosenberger and Gasko, 1983, p. 317), but it is an approximate solution that turns out to be more instructive.

Figure 7.1 shows the standard normal and mixed normal distributions considered here. Although the mixed normal is symmetric and bell-shaped, it is not a normal curve. A probability curve is normal only if the equation for the curve belongs to the family of curves given by Equation 3.1 in Chapter 3. That is, the mixed normal belongs to the family of normal curves if we can find a mean (μ) and standard deviation (σ) such that when plugged into Equation 3.1, we get the probability curve for the mixed normal exactly, and it can be verified that this cannot be done. This illustrates that if we assume all probability curves are normal, we get a contradiction. This assumption cannot be true because when we mix two distinct normal curves together, in general we do not get a normal curve back. Indeed, *there are infinitely many bell-shaped curves that are not normal.*

As is evident, there appears to be little difference between the normal and mixed normal curves in Figure 7.1. In fact, it can be shown that if we want to determine the probability that an adult's self-esteem score is less than c, for any c we might pick, we get a very good approximation by assuming that the mixed normal is in fact a normal distribution with mean zero and variance one. More precisely, when sampling an adult, if we assume that all self-esteem scores are standard normal when determining $P(X \leq c)$, the correct probability will not be off by more than .04. Put another way, mixing schizophrenics in with nonschizophrenics does not change $P(X \leq c)$ very much. (Mathematical statisticians have a formal method for describing this result using the so-called Kolmogorov distance between two probability curves.)

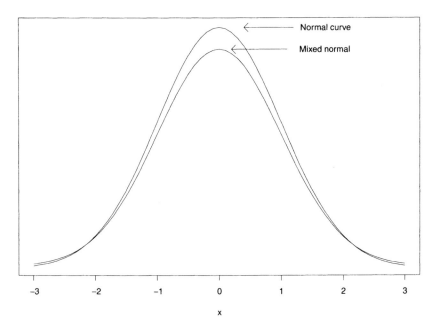

FIGURE 7.1 • A plot of the standard normal and mixed normal probability curves. Despite the similarity between the two curves, one has variance 1, and the other has variance 10.9.

It should be noted that the mixed normal just described is just one member of a family of probability curves that are called mixed normal. For example, rather than sample from a standard normal curve with probability .9, we could reset this probability to .95, .8, or any value between 0 and 1. Similarly, the standard deviation of the second normal curve need not be 10; it could be any positive number. Here we use the specific mixed normal considered by Tukey because it provides a simple way of illustrating general problems associated with nonnormality.

Tukey used the contaminated normal to argue that, in practice, outliers are common and should be expected, but he offered no empirical justification for this prediction. He could not because at the time, effective outlier detection methods had not yet been devised. Today such methods are widely available, and it turns out that Tukey was correct. This is not to say that outliers always appear, but they are certainly more common than one might guess, and they arise in situations where they might seem completely unexpected.

There are many important implications associated with the mixed normal that were not discussed by Tukey but are readily derived using basic theoretical principles. These practical concerns stem from the following important point that was discussed by Tukey. Despite the small difference between the normal and mixed normal curves just described, their variances differ substantially. The variance of the standard normal is $\sigma^2 = 1$, but the variance of the mixed normal is $\sigma^2 = 10.9$. This illustrates a general property of extreme importance: *The population variance is very sensitive to the tails of any probability curve.* Put another way, an extremely small subset of the population of people (or things) under study—individuals that are relatively rare and atypical—can completely dominate the value of the population variance. In fact, examples can be constructed where a probability curve is even more similar to the normal curve in Figure 7.1 than is the mixed normal considered here, yet the variance is even larger than 10.9. (Indeed, using a simple extension of the results in Staudte and Sheather, 1990, Section 3.2.1, the variance can be made arbitrarily large.) From a modern perspective, *the population variance is not robust*, roughly meaning that very small changes in *any* probability curve—not just the normal curve—can have a large impact on its value.

POWER

The lack of robustness associated with the population variance has devastating implications for a number of practical problems. One reason has to do with our ability to detect differences between two groups of subjects. Consider, for example, two independent groups, one of which receives a standard method for treating depression and the other is treated with an experimental method. Further assume that both groups have normal distributions with variance one, the first has mean zero, and the second has mean one, so the distributions appear as shown in Figure 7.2. If we sample twenty-five individuals from both groups and compare the means with Student's test (at the .05 level), then power, the probability of deciding there is a difference between the means, is about .96. That is, we have about a 96 percent chance of discovering that there is a difference between the means of the two groups and concluding that the new method for treating depression offers an advantage, on average, over the traditional approach. If we switch to the percentile t method using W, as described in Chapter 6, power is slightly lower.

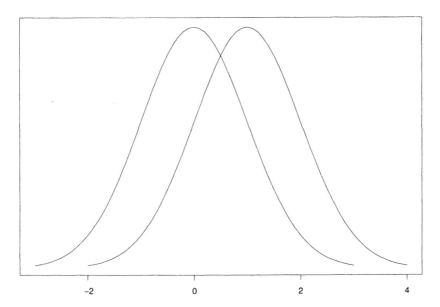

FIGURE 7.2 • Two normal curves with means 0 and 1. When sampling twenty-five observations from each curve, there is a 96 percent chance of detecting the difference between the means with Student's T when testing at the .05 level.

Now imagine that, unknown to us, both groups have the mixed normal probability curve previously described. That is, the curves for both groups now appear as shown in Figure 7.3, and again the difference between the means is one. As is evident, there is little visible difference from Figure 7.2, yet the probability of detecting the difference between the means is now only .28, and switching to the bootstrap method for means does not increase power. A discovery has a good chance of being missed under even an arbitrarily small departure from normality! The reason power is so low when sampling from the mixed normals is that when working with means, power is inversely related to the population variance, as noted in Chapter 4. In more formal terms, the squared standard error of the sample mean is

$$\text{VAR}(\bar{X}) = \frac{\sigma^2}{n},$$

which can become very large under very slight departures from normality. For *any* method based on the sample mean, as the population variance (σ^2) gets large, power decreases. So we see that if we assume normality, and this assumption is even slightly violated, the result can be a very serious decrease in power when comparing means.

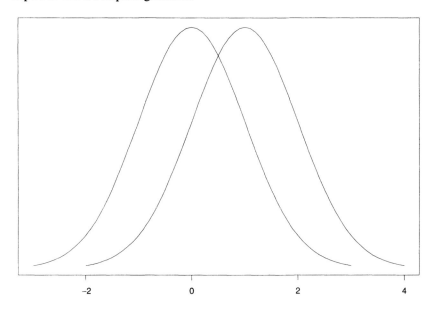

FIGURE 7.3 • Two mixed normals with means 0 and 1. Despite the similarity to Figure 7.2, there is now only a 28 percent chance of detecting the difference between the means. This illustrates that if the normality assumption is violated even slightly, the power of any method based on means might be lowered substantially.

Heavy-tailed distributions, such as the mixed normal, are characterized by outliers, so a reasonable speculation about how to handle the problem just illustrated is to check for outliers and to use means if none are found. Unfortunately, this approach can be highly unsatisfactory because other differences between probability curves, such as unequal variances or differences in skewness, can greatly affect our ability to detect true differences between means, as noted in Chapter 5. A better strategy, at least for the moment, is to use a method that performs about as well as methods based on means when normality is true, but it continues to work well in situations where methods based on means perform poorly. Efforts to find such methods have been successful and will be described in subsequent chapters.

As noted in Chapter 1, it is often argued that if a probability curve appears to be reasonably symmetric and bell-shaped, there is no need to worry about nonnormality when statistical methods based on the sample mean are employed. In fact, conventional wisdom holds that generally nonnormality is not an issue when comparing groups. This means that if two groups have *identical* probability curves, the probability of a Type I error (declaring a difference between the means when in fact none exists) can be controlled reasonably well. But the illustration just given demonstrates that if groups differ, nonnormality might be a serious concern because an important difference might be missed.

ANOTHER LOOK AT ACCURACY

Chapter 4 described the sampling distribution of both the mean and median. It was illustrated that for the normal curve the sample mean tends to be closer to the population mean than the median but for Laplace's probability curve, the reverse is true. Let's repeat our illustration, only now we sample observations from a mixed normal instead. If we sample twenty observations and compute both the mean and median and if we repeat this four thousand times and plot the results, we get the curves shown in Figure 7.4. As is evident, the median tends to be a much more accurate estimate of the central value (μ).

Figure 7.4 suggests that when we compare two groups of subjects having mixed normal probability curves, we have a higher probability of detecting a true difference (higher power) if we use medians rather than means, and this speculation is correct. It is not, however, being argued that the median be used routinely. The problem is that for light-tailed probability curves, such as the normal, the sample mean provides considerably more power than the median. For example, we previously described a situation where power is .96 when comparing the means of two normal distributions with Student's T. If we compare medians instead, power drops to .74. What would be useful is an estimator that provides reasonably high power when sampling from a normal curve, yet the probability of detecting a true difference remains high when sampling from a mixed normal.

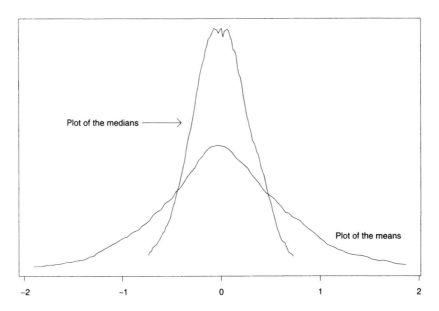

FIGURE 7.4 • A plot of means versus medians when sampling from a mixed normal. In this case the plot of the medians is more tightly centered around zero, the value being estimated, indicating that it tends to be a more accurate estimate of the central value.

THE GRAPHICAL INTERPRETATION OF VARIANCE

As noted in Chapter 3, there is a distinct and obvious difference between two normal probability curves having standard deviations 1 and 1.5. This might suggest that if two probability curves have both equal means and equal variances, then a graph of these two curves will be very similar, but this is not necessarily the case even when they are bell-shaped. Figure 7.5 shows two symmetric probability curves centered around the same value, so they have equal means. Both probability curves have equal variances, yet, as is evident, there is a clear and substantial difference between the two curves. One is a normal curve, but the other is the same mixed normal shown in Figure 7.1. Put another way, if unknown to us we sample observations from a mixed normal but we attempt to approximate the probability curve using a normal curve with the same mean and variance as the mixed normal, we get a poor approximation despite the fact that both curves are bell-shaped. If we allow the possibility that a probability curve might be skewed, there might be an even more striking difference between two probability curves, despite having equal means and variances, as illustrated by Figure 7.6.

The Graphical Interpretation of Variance

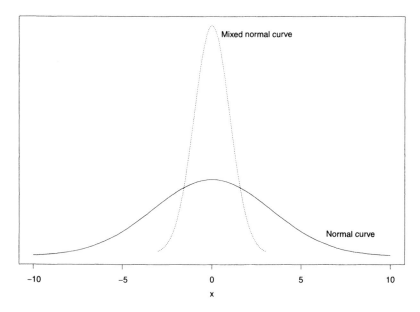

FIGURE 7.5 • An illustration that even when probability curves are symmetric and bell-shaped and have equal means and variances, the curves can differ substantially.

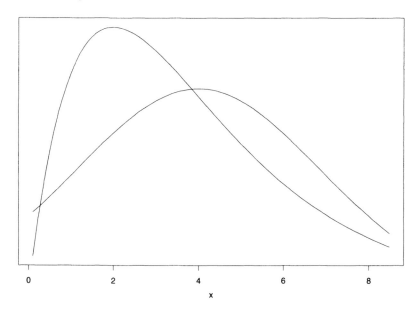

FIGURE 7.6 • Another illustration of how different two probability curves can be even when they have equal means and variances.

OUTLIER DETECTION

Chapter 3 described an outlier detection method motivated by a basic property of the normal curve: The probability of an observation being within two standard deviations of the mean is always .954. So if we declare an observation to be an outlier when it is more than two standard deviations from the mean, there is a .046 probability that an observation will be declared an outlier. It was mentioned in Chapter 3 that even when the standard deviation is known exactly, using it to detect outliers might result in missing outliers due to masking. The mixed normal illustrates the problem. As indicated in Figure 7.7, for the mixed normal considered here, the probability of an observation being within *one* standard deviation of the mean is .999 versus .68 for the normal probability curve. If we declared a value more than two standard deviations from the mean to be an outlier, then for the mixed normal the probability of finding an outlier is approximately 4.1×10^{-11}. That is, it is virtually impossible for an observation to be declared an outlier despite the similarity between the normal and mixed normal probability curves. Again, the variance is so sensitive to the tails of a probability curve that it can be highly unsatisfactory when used to detect outliers.

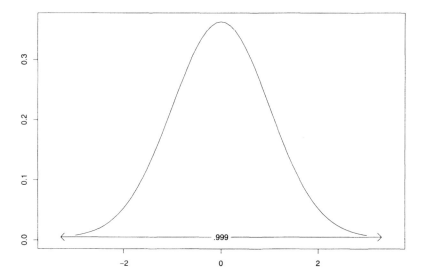

FIGURE 7.7 • For any normal curve, the probability of an observation being within one standard deviation of the mean is .68. But even for a small departure from normality, this probability can exceed .999, as illustrated here with the mixed normal.

MEASURING EFFECT SIZE

A fundamental problem is finding numerical methods to characterize the difference between two groups of subjects. An obvious approach is to use the difference between the means. Another frequently employed approach is to use a standardized difference. As in Chapter 5, let μ_1 and μ_2 be the population means of two groups with corresponding standard deviations σ_1 and σ_2. Momentarily assume that the standard deviations are equal. Letting σ represent this common value, a frequently used measure of effect size is the standardized difference

$$\Delta = \frac{\mu_1 - \mu_2}{\sigma}.$$

Jacob Cohen defined a large effect size as one visible to the naked eye when viewing the probability curves. When both groups have a normal curve, he concluded that $\Delta = .8$ is large, $\Delta = .2$ is small, and $\Delta = .5$ is a medium effect size. Figure 7.8 shows two normal curves, each having standard deviation 1, with means 0 and .8, respectively. So, $\Delta = .8$, which is considered a large difference. Now look at Figure 7.9. The difference between the means is the same as before, but the probability curves are contaminated normals with variance 10.9. (A difference between Figures 7.8 and 7.9 can be discerned by noting that in Figure 7.8 the y-axis extends to .4, but in Figure 7.9 it does not.) So now $\Delta = .24$, suggesting the difference is small, but the graphs of the probability curves indicate that the difference is large. That is, if we use means and variances to measure effect size, to the exclusion of all other tools we might employ, situations arise where we will underestimate the degree to which groups differ.

HOW EXTREME CAN THE MEAN BE?

For symmetric probability curves, such as the normal curve, there is a .5 probability that an observation is less than the population mean, μ. That is, the population mean lies at the center of the infinitely many observations if only they could be observed. But how small can this probability become? Put another way, how far into the tail of a skewed curve can the mean be? The answer is that the probability can be arbitrarily close to zero or one, and the population mean can be unlimitedly far into the tail. The proof of these statements are based on an approach that is similar in spirit to the mixed normal (e.g., Staudte and Sheather, 1990), but the details are too involved to give here.

128 CHAPTER 7 / *A Fundamental Problem*

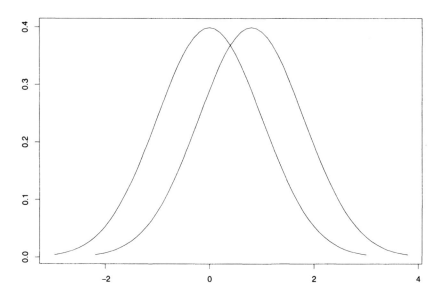

FIGURE 7.8 • Two normal probability curves for which $\Delta = .8$.

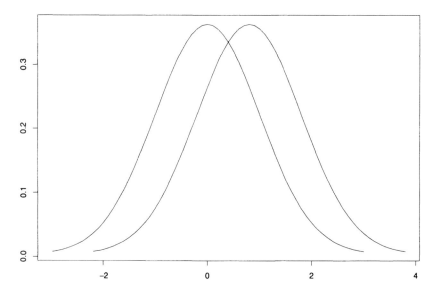

FIGURE 7.9 • Despite the similarity to Figure 7.8, $\Delta = .24$ because the curves are now mixed normals.

As many introductory books suggest, if a probability curve is too skewed and the population mean is in the extreme portion of the tail of a probability curve, the population mean is no longer a good measure of central tendency—it can give a distorted and misleading indication of the typical object or person under study. There is, however, no agreed-on criterion for judging the adequacy of the population mean. How far from the central portion of the data can the population mean be before we declare it to be an unsatisfactory summary of the data?

REGRESSION

When we consider regression (fitting a line to a scatterplot of points), the situation becomes more involved. To begin, we focus on the usual least squares estimate of the slope described in Chapter 2. Currently, this is the estimator routinely used. For illustrative purposes, let's again consider the study of diabetes in children where one specific goal is determining whether the age of a child at diagnosis could be used to predict a child's C-peptide concentrations. For the moment, attention is focused on the homoscedastic case. As explained in Chapter 4, this means that the conditional variance of C-peptide concentrations (Y) given a child's age (X), does not depend on X. For example, homoscedasticity implies that the population variance of the C-peptide concentrations, given that the child is 7 ($X = 7$), is equal to the variance for children who are 8, or any other age we pick. As before, we label this common variance σ^2.

Now consider the problem of testing the hypothesis that the slope is zero. There is a routinely used method for accomplishing this goal that was outlined in Chapter 5 and is based on the least squares estimate of the slope. (Virtually all introductory textbooks describe the details.) Naturally, if the true slope differs from zero, we want a high probability of detecting this. That is, we want the power of any test we use to be reasonably high. If we compute, say, a .95 confidence interval for the slope, we want the length of the confidence interval to be as short as possible. But to keep the discussion relatively simple, we focus on the goal of maintaining high power.

Power depends on the squared standard error of the estimator being used. As indicated in Chapter 4, when using the least squares estimator of the slope, the squared standard error is

$$\text{VAR}(b_1) = \frac{\sigma^2}{(n-1)s_x^2}.$$

So if, for example, the C-peptide levels have a contaminated normal distribution rather than a normal, this has a large impact on the variance, σ^2, which in turn means that our ability to detect a nonzero slope is relatively small compared to what it would be under normality. That is, again a small departure from normality can substantially reduce our ability to detect a true association.

Note, however, that outliers among the X values, called *leverage points*, will inflate the sample variance s_x^2, as noted in Chapter 2, and this will decrease the standard error of the least squares estimate of the slope. This suggests that leverage points are beneficial in terms of increasing our ability to detect regression lines having a nonzero slope, but there is yet another consideration that must be taken into account. If the X value is a leverage point, and simultaneously the corresponding Y value is an outlier, we have what is called a *regression outlier* that might completely distort how the bulk of the points are related. That is, we might reject the hypothesis that the slope is zero and conclude that it differs substantially from zero even when it does not. In a similar manner, we might fail to detect a situation where the slope differs from zero, not because the slope is indeed zero, but because regression outliers mask an association among the bulk of the points under study.

We illustrate this last point using data from an actual study. Figure 7.10 shows a scatterplot of points where the goal is to study predictors of reading ability in children. (The data were collected by L. Doi.) Also shown is the least squares regression line, which has an estimated slope of −0.032. Using the method in Section 6.5 of Chapter 6 for computing a .95 confidence interval for the true slope yields (−0.074, 0.138). This interval contains zero, and therefore we are unable to conclude that the particular variable under study predicts reading ability. In fact, the hypothesis of a zero slope would not be rejected even if he had computed a .5 confidence interval. (That is, if we set the probability of a Type I error at .5, we still do not reject.) So as suggested by Figure 7.10, the least squares regression line offers no hint of an association, and it might seem that this conclusion is surely correct.

Now look at Figure 7.11, which is based on the same data shown in Figure 7.10, with two features added. The first is a relplot (derived by K. Goldberg and B. Iglewicz), which is distinguished by the two ellipses; it's a bivariate analog of the boxplot covered in Chapter 2. The inner ellipse contains the middle half of the data, and points outside the outer ellipse are declared outliers. So the six isolated points in the right portion of Figure 7.11 are labeled outliers. The second feature is the ragged line in Figure 7.11, which is called a *running interval smoother*. It represents an approximation of the regression line without forcing it to have a particular shape such as a straight line. As is

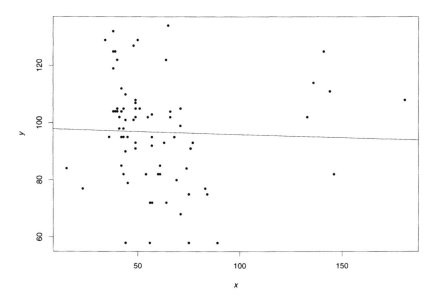

FIGURE 7.10 • A scatterplot of the reading data. The nearly horizontal line in the center of the graph is the least squares regression line.

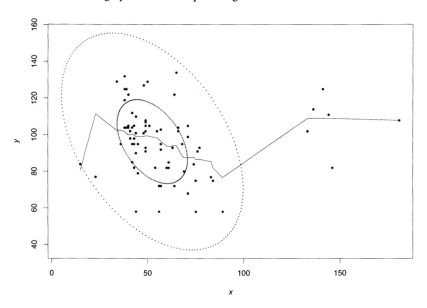

FIGURE 7.11 • Another scatterplot of the reading data used in Figure 7.10 with a relplot and smooth added. This illustrates that modern graphical methods can provide a substantially different perspective than more commonly used techniques.

evident, Figure 7.11 suggests that for the bulk of the observations, there is a negative association between the two variables under study. Using any one of several methods described in Part II of this book, this negative association can be confirmed. It is missed by the least squares regression line because of the outliers. In fact, a single outlier can cause problems because the least squares regression line has a finite sample breakdown point of only $1/n$. That is, a single unusual point can cause the slope to take on any value regardless of the association among the remaining points.

PEARSON'S CORRELATION

Chapter 6 introduced Pearson's correlation ρ and noted that the estimate of ρ, r, has a finite sample breakdown point of only $1/n$. One of the main disadvantages of this property is that associations can be missed that are detected by more modern techniques. To illustrate that restricting attention to Pearson's correlation can have practical consequences, consider again the data in Figure 7.10. We get $r = -0.03$, and applying Student's T test of $H_0 : \rho = 0$, we fail to reject at the .05 level (the significance level is .76). So, consistent with our result based on the least squares estimate of the slope, no association is found, but several techniques developed in recent years indicate that there is indeed an association. (Some of these methods will be described in Chapters 10 and 11.) Switching to the modified percentile bootstrap method described in Chapter 6, we still fail to reject $H_0 : \rho = 0$, but employing the bootstrap does not correct problems due to nonnormality because these problems are intrinsic to ρ and its estimate, r.

Even if we know the value of ρ exactly, its value can be extremely misleading. For example, the left panel of Figure 7.12 shows a plot of the joint probability curve for X and Y when both are normal with a correlation of .8. The right panel shows the same situation with the correlation $\rho = .2$. So under normality, we have some sense of how a probability curve changes with ρ. But now look at Figure 7.13. Again the correlation is .2, but the probability curve looks like the left panel of Figure 7.12, where $\rho = .8$. In Figure 7.13, X again has a normal probability curve, but Y has a mixed normal curve instead. So we see that a very small change in one of the (marginal) probability curves can have a large impact on the value of Pearson's correlation coefficient, ρ.

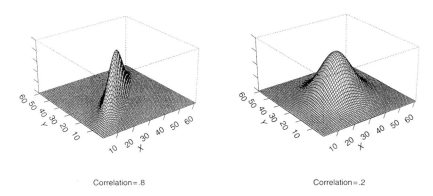

FIGURE 7.12 • An illustration of how ρ alters a bivariate normal probability curve.

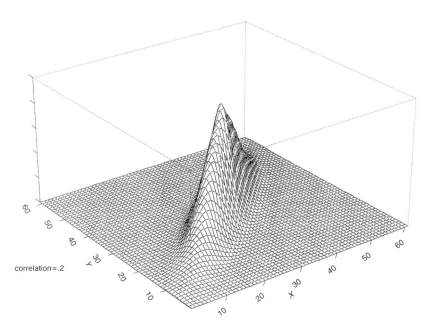

FIGURE 7.13 • This bivariate probability curve is similar to the curve in the left panel of Figure 7.12, but the correlation is $\rho = .2$. Here X is normal but Y has a mixed normal probability curve. This illustrates that small departures from normality can drastically alter ρ.

MORE ABOUT OUTLIER DETECTION

Before ending this chapter, some remarks about detecting outliers might be useful. When working with bivariate data, such as shown in Figure 7.11, a natural strategy for detecting outliers is to check a boxplot of the X values and then do the same for the Y values. There is, however, a problem. Look at Figure 7.11 and rotate this book. That is, begin by holding the book in the usual upright position and then rotate it clockwise. As you rotate the book, the outliers in Figure 7.11 should remain outliers. That is, the angle at which you view a scatterplot of points should not matter when declaring a point to be an outlier. It turns out that if we rotate the points, while leaving the axes fixed, simply applying a boxplot to each of the two variables might give different results depending on how much we rotate the points. Some points might be declared an outlier when holding this book upright but not when it is rotated say 45 degrees. Suffice it to say that there are methods for dealing with this issue; one of them is the relplot in Figure 7.11.

A SUMMARY OF KEY POINTS

- Small departures from normality can inflate σ tremendously. In practical terms, small departures from normality can mean low power and relatively long confidence intervals when working with means.
- Approximating a probability curve with a normal curve can be highly inaccurate, even when the probability curve is symmetric.
- Even when σ is known, outlier detection rules based on $\bar{X}$ and σ can suffer from masking.
- The standardized difference, Δ, can grossly underestimate effect size when dealing with nonnormal distributions.
- Leverage points can reduce the standard error of the least squares estimate of a slope, but generally outliers among the scatterplot of points can cause the least squares estimator to give a distorted and misleading summary of data. Also, outliers can inflate the squared standard error of the least squares estimator.

- Special methods for detecting outliers among multivariate data are needed. Simply using a boxplot on the X values and doing the same with the Y values can be unsatisfactory.

BIBLIOGRAPHIC NOTES

There are many ways of estimating the shape of a regression line without assuming it is straight or has some specific form. For a detailed discussion of these methods, see Hastie and Tibshirani (1990). More details about the running interval smoother in Figure 7.11 can be found in Wilcox (1997). For a description of how poorly the standard confidence interval for the slope can perform, as well as the improvement afforded by the modified percentile bootstrap described here, see Wilcox (1996a). Tukey (1960) discussed the contaminated normal; his paper marks the beginning of a series of serious crises regarding standard methods for making inferences about means and a least squares regression line. For a formal description of the problem of detecting outliers when points are rotated, see Rousseeuw and Leroy (1987). For a book devoted to outlier detection, see Barnett and Lewis (1994). For Cohen's discussion on how to use Δ to measure effect size, see Cohen (1977). For details about the relplot in Figure 7.11, see Goldberg and Iglewicz (1992).

PART TWO

CHAPTER 8

ROBUST MEASURES OF LOCATION

Prior to the year 1960, there was a very poor understanding of how to study nonnormality and its effects on conventional hypothesis testing methods. Limited studies seemed to suggest that nonnormality is not an issue when the goal is to make inferences about population means. But starting in 1960, a series of crises made it clear that finding improved methods is critical to those conducting applied research. These crises can be roughly classified into three groups, with a myriad of details in each.

The first group (discussed in Chapter 7) has to do with the realization that arbitrarily small departures from normality can have devastating consequences on the population mean, particularly the population variance. One consequence is that nonnormality can result in very low power and an extremely poor assessment of effect size when attention is restricted to means and variances. Problems get worse when trying to detect and measure the association between variables via least squares regression and Pearson's correlation. A fundamental problem is that both the population mean and the population variance are not robust, meaning that their values can be extremely sensitive to very small changes in any probability curve.

The second group has to do with a better understanding of how differences between probability curves can affect conventional hypothesis testing methods such as Student's T and its generalization to multiple groups using the so-called ANOVA F test. The first relevant study was conducted by George Box in 1954 and described situations where, when sampling from normal distributions, having unequal variances has no serious impact on the probability of a Type I error. Box's paper was mathematically sophisticated for its time, but it would be another ten years before mathematicians began to realize where we should look for serious practical problems. Through modern eyes, Box's numerical results were restricted to a rather narrow range of situations. When comparing groups, he considered instances where the largest of the standard deviations (among the groups being compared), divided by the smallest standard deviation, is less than or equal to $\sqrt{3}$. Even under normality, if we allow this ratio to be a bit larger, practical problems begin to emerge. In 1972, Gene Glass and two of his colleagues published a paper indicating that standard methods (Student's T and the ANOVA F test) are adversely affected by unequal variances and should be abandoned. With the advent of high-speed computers, new and more serious problems were found. By 1978, some quantitative experts began to call into question all applied research because of our inability to control errors with conventional methods. During the 1980s, statistical experts found even more devastating evidence for being unsatisfied with standard methods. For example, if groups differ in terms of both variances and skewness, we get unsatisfactory power properties (biased tests), a result mentioned in Chapter 5. In 1998, H. Keselman and some of his colleagues published another review paper summarizing problems and calling for the use of more modern techniques. When attention is turned to regression, these practical problems are exacerbated.

The third group in our tripartite collection of crises is the apparent erosion in the lines of communication between mathematical statisticians and applied researchers. Despite hundreds of journal articles describing problems with standard methods, the issues remain relatively unknown—and certainly underappreciated—among most applied researchers busy keeping up with their own areas of research. There has been such an explosion of developments in statistics that it is difficult for even statisticians to keep up with all the important issues. Moreover, quick explanations of more modern methods are difficult—maybe even impossible—for someone whose main business is outside the realm of mathematical statistics. Some modern methods are not intuitive based on the standard training most applied researchers receive, and with limited technical training they might appear to be extremely unreasonable. So the goal in the remaining chapters is to attempt to bridge this gap.

We need to divide and conquer the problems that have been described. We begin by taking up the problem of finding better estimators of central tendency—methods that offer practical advantages over the mean and median. What would such an estimator be like? Its standard error should not be grossly affected when sampling from distributions that represent only a very slight departure from normality. In terms of hypothesis testing, we want power to be nearly as high when using means and sampling is from a normal probability curve. But unlike the mean, we want power to remain relatively high when sampling from a heavy-tailed probability curve such as the mixed normal. Said another way, when using an estimator of location, its squared standard error (the variance of the estimator) should compete relatively well with the sample mean when we are sampling from a normal curve, but the squared standard error should not be grossly affected by small departures from normality. As we have seen, the sample median does not satisfy the first criterion, so we must look elsewhere based on the two conditions just described.

Another criterion is that the value of our measure of central tendency should not be overly sensitive to very minute changes in a probability curve. If, for example, we sample from a normal curve with a population mean of 10, then the population mean is a reasonable measure of the typical individual or thing being studied. It is possible to alter this curve so that it still appears to be approximately normal, yet the population mean is increased by a substantial amount. (Details can be found in books cited at the end of this chapter.) A related issue is that the population mean can be far into the tail of a skewed probability curve. At some point doubt arises as to whether the population mean provides a reasonable measure of what is typical. What would be nice is some assurance that regardless of how skewed a probability curve happens to be, our measure of central tendency will be near the bulk of the most likely values we might observe when sampling observations.

THE TRIMMED MEAN

Currently, there are two classes of estimators that satisfy the criteria just described and simultaneously have practical value when testing hypotheses. The first is called a *trimmed mean* and is discussed in detail here. The second is called an M-estimator and will be described later in this chapter.

In Chapter 2 it was noted that the sample median can be viewed as a type of trimmed mean: All but the middle one or two observations are removed and the remaining observations are averaged. In contrast is the sample mean, which involves no trimming at all. So in terms of how much to trim, these two estimators represent extremes, both of which have problems we've already discussed. The idea behind the class of trimmed means is to use a compromise amount of trimming. For example, if we observe the values

$$2, 5, 7, 14, 18, 25, 42,$$

the median is 14. That is, we trim the three smallest and three largest values to get the median. But we could have trimmed just the smallest and largest values. If we do this and average the five remaining values, we have what is called a *trimmed mean*, which will be labeled $\bar{X}_t$. Carrying out the computations for the data at hand, we get

$$\bar{X}_t = \frac{1}{5}(5 + 7 + 14 + 18 + 25) = 13.8.$$

This type of trimming is routinely used in certain sporting events, such as Olympic figure skating. Each skater is rated by a panel of judges, and the contestant's score is based on the average rating after the highest and lowest ratings are removed. That is, a trimmed mean is used, the idea being to ignore ratings that might be atypical. But for our purposes, there are a myriad of technical issues that must be addressed. Among these is deciding how much to trim. In our example, rather than remove the largest and smallest values, why not remove the two smallest and two largest values instead? Now our trimmed mean is

$$\bar{X}_t = \frac{1}{3}(7 + 14 + 18) = 13.$$

Not surprisingly we get a different result from the first trimmed mean, so the relative merits associated with different amounts of trimming need to be considered.

We begin our discussion of this important issue by considering the standard normal versus the mixed normal described in Chapter 7. We saw that the variance of the sample mean is considerably smaller than the variance of the sample median when sampling from a normal curve, but the reverse is true when sampling from the mixed normal instead, which represents a very slight departure from normality. A natural strategy is to consider how much we can trim and still compete reasonably well with the sample mean when sampling

The Trimmed Mean

from a normal curve and then consider whether this amount of trimming competes reasonably well with the median when sampling from the mixed normal instead. In effect, the idea is to guard against disaster. We cannot find an estimator that is always optimal—no such estimator exists—so we do the next best thing and look for an amount of trimming that beats both the mean and median by a considerable amount in some cases, and simultaneously there are no situations where the reverse is true.

Based on this strategy, it has been found that 20 percent trimming is a good choice for general use. This means that we proceed as follows. Determine n, the number of observations, compute $.2n$, and then round down to the nearest integer. For example, if $n = 19$, $.2n = 3.8$, and when we round down 3.8 to the nearest integer we get 3. Then the trimmed mean is computed by removing the three smallest and three largest of the 19 values and averaging those that remain. More generally, if when we round $.2n$ down to the nearest integer we get the integer g, say, remove the g smallest and largest values and average the $n - 2g$ values that remain.

To graphically illustrate how the 20 percent trimmed mean compares to the sample mean when sampling observations from the mixed normal, we repeat our computer experiment from Chapter 7 (the results of which were shown in Figure 7.4) with the median replaced by the 20 percent trimmed mean. That is, we generate twenty observations from a mixed normal, compute the mean and 20 percent trimmed mean, repeat this four thousand times, and then plot the resulting means and trimmed means. Figure 8.1 shows the results. As indicated, the 20 percent trimmed mean tends to be substantially closer to the central value zero. That is, when sampling from a symmetric probability curve, both estimators are attempting to estimate the population mean, which in this case is zero. So the estimator that tends to be closer to zero is better on average, and here this is the 20 percent trimmed mean. If we sample from the standard normal curve instead, theory indicates that the sample mean will tend to be more accurate. This is illustrated by Figure 8.2, but the increased accuracy is not very striking.

Notice that a trimmed mean is not a weighted mean. The trimmed mean involves ordering the observations, which puts us outside the class of weighted means covered by the Gauss-Markov theorem described in Chapter 4.

We saw in Chapter 7 that the median can be substantially more accurate than the mean, so it is not completely surprising that the trimmed mean can be substantially more accurate. But it is important to develop a reasonable amount of intuition as to why this is. Despite illustrations about how poorly the sample mean can perform compared to the trimmed mean, there is often

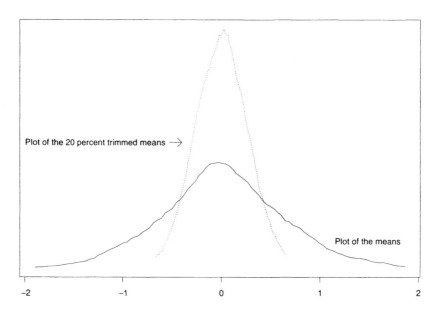

FIGURE 8.1 • Like the median, the 20 percent trimmed mean is a substantially more accurate estimator than the mean when sampling from a mixed normal probability curve.

a reluctance to use a trimmed mean because the idea that discarding data can result in increased accuracy is often viewed as being counterintuitive.

Perhaps the simplest explanation is based on the expression for the variance of the sample mean, which, as we have seen, is σ^2/n. From Chapter 7, arbitrarily small departures from normality can inflate the population variance, σ^2, tremendously. That is, the variance of the sample mean is extremely sensitive to the tails of a probability curve. By trimming, we in effect remove the tails. That is, we remove the values that tend to be far from the center of the probability curve because these values can cause the sample mean to be highly inaccurate. In fact, from a certain perspective, it is not surprising that the trimmed mean beats the mean, but it is somewhat amazing that the sample mean ever beats a trimmed mean (or median) in terms of its variance (or squared standard error).

To elaborate on this last statement, consider a standard normal probability curve (having a population mean $\mu = 0$ and standard deviation $\sigma = 1$) and imagine that we randomly sample twenty observations. Next, focus on the smallest of these twenty values. It can be shown that with probability .983,

THE TRIMMED MEAN

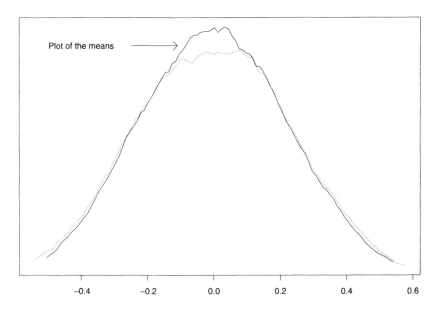

FIGURE 8.2 • When sampling from a normal curve, the sample mean is more accurate than the 20 percent trimmed mean, on average, but the increased accuracy is not very striking.

the smallest value will be less than −0.9. That is, with a fairly high probability, it will be relatively far from the population mean of zero, the value we are trying to estimate based on our random sample of twenty observations. Briefly, the reason is that when randomly sampling a *single* observation from a standard normal curve, with probability .816 it will be greater than −0.9. But in order for the smallest of twenty observations to be greater than −0.9, all twenty must be greater than −0.9, and this probability is $.816^{20} = .017$. So the probability that the smallest value is less than −0.9 is 1 − .017 = .983. In a similar manner, the largest of the twenty values will be greater than .9 with probability .983. Moreover, if we increase the sample size, we can be even more certain that the smallest and largest values will be far from the population mean.

In contrast, the middle values among a random sample of observations are much more likely to be close to the center of the normal curve (the population mean). For example, still assuming normality and if we again randomly sample $n = 20$ observations, then with probability .95 the average of the two central values (the sample median), will be between −0.51 and 0.51. That is, the middle values are more likely to be close to the population mean, the thing we are trying to estimate, than the smallest and largest values. Yet the sample mean gives the extreme values the same amount of weight ($1/n$) as the middle values. From the perspective just described, a natural guess is that the observed values in the middle should be given more credence than those in the tails because we know that the values in the tails are unlikely to provide accurate information about the population mean. Yet, despite these properties, the sample mean beats both the median and trimmed mean in accuracy when sampling from a normal curve because when we remove extreme values from our data, the remaining observations are dependent. (The reason the observations are dependent may not be immediately obvious; a detailed explanation is postponed until Chapter 9, where this issue will take on an even more central role. Also see Appendix A.) It turns out that this dependence can tip the balance toward preferring the mean over the trimmed mean or median provided we sample from a normal probability curve or a curve with relatively light tails (meaning that outliers are rare). But for the very common situation where we sample from a heavy-tailed probability curve, the reverse can happen to an extreme degree, as was illustrated in Figure 8.1.

An objection to this simplified explanation is that when sampling from a symmetric curve, the smallest value will tend to be less than the population mean and the largest value will be greater than the population mean, so one could argue that if we include them when we compute the sample mean, they will tend to cancel each other out. Indeed, if we average the smallest and largest values, the expected value of this average will be μ. But again, it can be mathematically demonstrated that despite this, these extreme observations hurt more than they help under arbitrarily small departures from normality. More formally, extreme values can result in a relatively large standard error for $\bar{X}$.

So far the focus has been on sampling from a symmetric probability curve. Does sampling from a skewed curve change matters? In terms of getting an estimator with a relatively small variance, the answer is no. Regardless of whether we sample from a symmetric or skewed probability curve, the 20 percent trimmed mean will have a smaller variance when the curve is heavy-tailed, often by a substantial amount. One implication is that when testing hypotheses, we are more likely to get high power and relatively short confidence intervals if we use a trimmed mean than if we use the mean. Again the reason is that, regardless of whether a probability curve is skewed or symmetric, heavy-tailed distributions mean that the population variance will be inflated compared to situations where we sample from a light-tailed distribution. By eliminating the tail of a heavy-tailed probability curve, meaning that we trim, we avoid this problem and get an estimator with a relatively small variance. For a light-tailed curve, the sample mean might have a variance comparable to the 20 percent trimmed mean, but generally the improvement is not very striking. In Chapter 9 we will see that for skewed, light-tailed probability curves, trimming offers yet another advantage when testing hypotheses or computing confidence intervals.

Chapter 2 introduced the notion of a finite sample breakdown point and noted that the finite sample breakdown point of the sample mean is $1/n$ versus .5 for the median. The finite sample breakdown point of the 20 percent trimmed mean is .2, and with 10 percent trimming it is .1. So the minimum proportion of outliers required to make the 20 percent trimmed mean arbitrarily large is .2. Arguments have been made that a breakdown point less than or equal to .1 is dangerous. This is certainly true in terms of getting a relatively small standard error, and it applies when we want to avoid an estimate that is highly susceptible to outliers, in which case it might give us a distorted view of what is typical. When we turn to hypothesis testing, again we will see that there are advantages to avoiding a low finite sample breakdown point.

A natural reaction is that if a probability curve appears to be skewed to the right, say, trim some proportion of the largest observations but none of the smallest. The idea is that large outliers (outliers among the largest observed values) might arise, but outliers in the lower tail will tend to be rare. If the curve is skewed to the left, do the opposite. It has been found, however, that in the context of hypothesis testing, this strategy is less satisfactory, in terms of Type I errors and probability coverage, than always trimming both tails.

THE POPULATION TRIMMED MEAN

Like the mean and median, there is a population analog of the sample trimmed mean. For current purposes, it suffices to think of it as the value for the sample trimmed mean we would get if all individuals under study could be measured. With 20 percent trimming, for example, you would remove the lower 20 percent of all measures; if the entire population of individuals under study could be measured, do the same with the upper 20 percent, and then average the values that remain. For symmetric probability curves, the population mean, median, and trimmed mean are identical. But for a skewed probability curve, all three generally differ. In Figure 8.3, for example, the median and 20 percent trimmed mean have values that are near the most likely outcomes, but the population mean does not. For this reason, the median and 20 percent trimmed mean are often argued to be better measures of what is typical when distributions are markedly skewed.

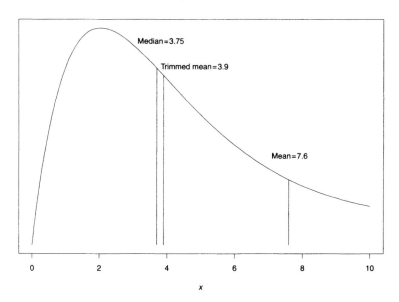

FIGURE 8.3 • For skewed probability curves, the population mean can have an atypical value. That is, it can lie in the extreme portion of the curve while the population median and trimmed mean have values that better reflect what we are more likely to observe when sampling an observation.

This is not to suggest, however, that outliers are always uninteresting. Clearly they can be very interesting and perhaps even the main focus of some investigation. But characterizing how groups differ in terms of outliers can be difficult. For one, such observations can be relatively rare, which in turn can make meaningful comparisons based on the outliers only a nontrivial task. One might argue that in some cases the population mean remains interesting when a probability curve is skewed with outliers because the goal is to capture how these outliers affect the probability curve. As long as one is clear about the interpretation of the population mean and the inherent problems with computing confidence intervals and testing hypotheses, perhaps there are situations where this view has merit.

One of the interesting theoretical developments since 1960 is a formal mathematical set of tools for characterizing the sensitivity of a parameter (such as the population mean or median) to small perturbations in a probability curve. The complete details go beyond the scope of this book, but we mention that three criteria form the foundation of modern robust methods: *qualitative robustness, infinitesimal robustness,* and *quantitative robustness.* The last criterion is related to the finite sample breakdown point of an estimator. Although it is *not* the formal definition used by theorists, we can think of quantitative robustness as the limiting value of the finite sample breakdown point. For example, the sample mean has a finite sample breakdown point of $1/n$, and this goes to zero as the sample size (n) goes to infinity. This limiting value is equal to what is called the *breakdown point* of the population mean. (The breakdown point is a numerical measure of quantitative robustness.) This means that we can alter a probability curve so that probabilities change by a very small amount but the population mean can be made arbitrarily small or large. In contrast, the population 20 percent trimmed mean and median have breakdown points of .2 and .5, respectively. A crude explanation is that about 20 percent of a probability curve must be altered to make the 20 percent population trimmed mean arbitrarily large or small. As for the median, about 50 percent of a probability curve must be altered.

The other two criteria are a bit more difficult to explain, but suffice it to say that the spirit of these criteria is to measure how small changes in a probability curve affect the values of population parameters. (These two criteria are related to generalizations of the notion of continuity and differentiability encountered in a basic calculus course.) The only important point here is that formal mathematical methods have been derived to judge measures of location and scale. By all three criteria, the population mean is the least satisfactory compared to any trimmed mean or the median.

M-ESTIMATORS

The other approach to measuring location that currently has considerable practical value consists of what are called *M*-estimators. Complete theoretical details motivating this approach are impossible here, but hopefully some sense of the thinking that led to interest in *M*-estimators can be conveyed.

To begin, recall from Chapter 2 that depending on how we measure error, we get different measures of location. For example, imagine we observe the values

3, 4, 8, 16, 24, 53.

As in Chapter 2, consider the goal of choosing a value c that is close to the six values just listed. If one prefers, we can view c as a value intended to predict an observed outcome. For example, if we choose $c = 4$, we get a perfect prediction for the second observation, but not for the others. If we measure closeness or the accuracy of our prediction, using the sum of the squared distances from c, this means we want to choose c to minimize

$$(3-c)^2 + (4-c)^2 + (8-c)^2 + (16-c)^2 + (24-c)^2 + (53-c)^2.$$

It follows that c must have the property that

$$(3-c) + (4-c) + (8-c) + (16-c) + (24-c) + (53-c) = 0, \quad (8.1)$$

from which we see that c is the sample mean, $\bar{X}$. Here, $\bar{X} = 18$. So if we measure how well c predicts the six observed values using squared error, the optimal choice for c is 18. But as shown by Laplace, if we replace squared error with absolute error, the optimal choice for c is the median.

To make progress, we need to look closer at the result derived by Ellis in 1844, which was briefly mentioned in Chapter 2. There are infinitely many ways we can measure how close c is to a collection of numbers. Different methods lead to different measures of location. What Ellis did was to characterize a large class of functions that lead to reasonable measures of location. To describe them, let $\Psi(x)$ be any function of x with the property that $\Psi(-x) = -\Psi(x)$. Such functions are said to be *odd*. One such odd function is $\Psi(x) = x$. For the six observations being considered here, consider the strategy of choosing our measure of location such that

$$\Psi(3-c) + \Psi(4-c) + \Psi(8-c) + \Psi(16-c) + \Psi(24-c) + \Psi(53-c) = 0. \quad (8.2)$$

For $\Psi(x) = x$, in which case $\Psi(x - c) = x - c$; this last equation reduces to Equation 8.1. That is, least squares and the sample mean correspond to choosing Ψ to be a straight line through the origin with a slope of one.

Another choice for Ψ is $\Psi(x) = \text{sign}(x)$, where $\text{sign}(x)$ is equal to $-1, 0,$ or 1 according to whether x is less than, equal to, or greater than zero. That is, negative numbers have a sign of -1, 0 has a sign of 0, and positive numbers have a sign of 1. This choice for Ψ leads to taking c to be the median. But we know that the median does not perform well under normality, and the mean does not perform well under arbitrary small departures from normality. So how do we choose Ψ to deal with these two problems?

Before answering this question, it helps to look at graphs of some choices for Ψ. Figure 8.4 shows four choices that have been considered by mathematicians. (Many others have also been considered.) The two top graphs

M-ESTIMATORS

correspond to using the mean and median. The graph in the lower-left portion of Figure 8.4 is an example of what is called Huber's Ψ. Generally, this class of Ψ functions is identical to the least squares Ψ for values of x between $-K$ and K, where K is some constant to be determined. That is, $\Psi(x) = x$ for $-K \leq x \leq K$. For x outside this range, Ψ becomes a horizontal line. In Figure 8.4, Huber's Ψ with $K = 1.28$ is shown. That is, the graph is exactly like the Ψ that leads to the mean provided x is not too far from the origin.

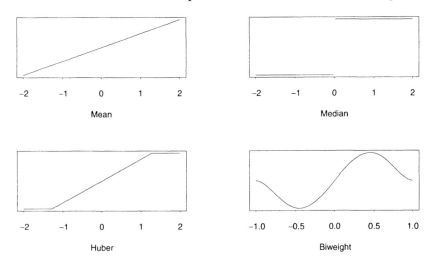

FIGURE 8.4 • Examples of Ψ functions that have been considered in connection with M-estimators of location.

In essence, if we use Huber's Ψ, we use least squares if an observation is not too extreme, but if an observation is extreme, it is down-weighted and possibly ignored. Complete details are impossible at an elementary level, but it can be seen that this strategy allows us to choose K so that the resulting estimate of location competes well with the mean under normality; we get good results when, for example, sampling is from a probability curve having heavier tails, such as the mixed normal described in Chapter 7. Notice that the graph in the upper-right portion of Figure 8.4 does not consist of a straight line through the origin. Rather, there is a sharp jump. The result is that the median does not behave like the mean when sampling from a normal curve. The Ψ function shown in the lower-right corner of Figure 8.4 is the so-called *biweight*.

The sample mean performs so poorly under nonnormality for reasons that can be related to the fact that the choice $\Psi(x) = x$ is unbounded. That is, as x gets large, there is no limit to how large $\Psi(x)$ becomes, and of course as x approaches minus infinity, so does $\Psi(x) = x$. It can be shown that as a result extreme observations can have an inordinately high influence on the value of the population mean, and by implication there are practical problems with the sample mean as well.

To say this in another manner, temporarily consider the case where observations are symmetric about the value c. In this case c corresponds to the population mean, median, and trimmed mean, and any reasonable choice for a measure of location would be this central value. Next, look again at Equation 8.2 and notice that if we use $\Psi(x - c) = x - c$, the further a value is from the center, the larger Ψ will be. It turns out that the difference $x - c$ (the distance a value x happens to be from the central value, c) reflects the influence the value x has on our measure of location. The same is true for skewed probability curves, but this is harder to explain simply. To avoid problems with nonnormality, we need to choose Ψ so that there are limits on how large or small its value can be. That is, we need to limit the influence of extreme values. Huber's Ψ is one such choice, where its value never exceeds K or drops below $-K$. By bounding Ψ, we limit how much an extreme value can influence our measure of location.

But there are infinitely many choices for Ψ that are bounded. Why not use the biweight shown in the lower-right portion of Figure 8.4? The biweight is an example of what is called a *redescending* Ψ. But all such Ψ have a technical problem that eliminates them from consideration. (The problem is related to how a measure of location is estimated based on observations available to us.) Huber's Ψ avoids this problem and is currently one of the more interesting choices from both a practical and technical view.

One more problem needs to be described. Imagine we want to find a number that characterizes the typical height of an adult living in France. If according to our measure of central location the typical height is 5 feet, 8 inches, then this is equivalent to 68 inches. The point is that when we multiply all observations by some constant, such as multiplying feet by 12 to get inches, the measure of location we are using should also be multiplied by this constant. Such measures of location are said to be *scale equivariant*. But for a wide range of Ψ values, we do not get this property automatically—a measure of scale must be incorporated into our Ψ function to achieve scale equivariance. A useful and effective measure of scale is the median absolute deviation statistic, MAD, which was introduced in Chapter 3. One basic reason is that

MAD has a breakdown point of .5 and this translates into a breakdown point of .5 when using an M-estimator of location with Huber's Ψ. For the data considered here, the goal is to use as our measure of location the value c satisfying

$$\Psi\left(\frac{3-c}{MAD}\right) + \cdots + \Psi\left(\frac{53-c}{MAD}\right) = 0. \tag{8.3}$$

Finally, there is the problem of choosing K when using Huber's Ψ, and arguments for setting $K = 1.28$ have been made by quantitative experts. One reason is that the resulting variance of the M-estimator will be nearly comparable to the sample mean when sampling from a normal curve, and it will have a much smaller variance than the sample mean when sampling observations from a heavy-tailed probability curve.

COMPUTING A ONE-STEP M-ESTIMATOR OF LOCATION

We have suggested that an M-estimator with Huber's Ψ is of interest from an applied point of view, but how do we calculate it based on observations we make? There are two (closely related) strategies one might employ; all indications are that the more complicated method offers no practical advantage over the simpler one so only the simpler method is covered here. Called a one-step M-estimator (meaning that we use one iteration in what mathematicians call the Newton-Raphson method), we begin by empirically determining which values, if any, are outliers based on the sample median, M, and the measure of scale, MAD. We have already seen this method in Chapter 3, but for current purposes a slight modification turns out to be useful. Here, any observed value, X, is declared an outlier if

$$\frac{|X - M|}{MAD/.6745} > 1.28, \tag{8.4}$$

where the value 1.28 corresponds to our choice for K in Huber's Ψ. (In Chapter 3, the value 2 was used rather than 1.28 when detecting outliers.) As a simple illustration, consider again the values 3, 4, 8, 16, 24, and 53. Then $M = 12$, $MAD/.6745 = 12.6$, from which we find that 53 is declared an outlier. Let L be the number of outliers smaller than the median. In our example there are no such outliers, so $L = 0$. Similarly, let U be the number of outliers greater than the median; here, $U = 1$. Next, sum the values that are not labeled outliers and call it B. In our illustration, $B = 3+4+8+16+24 = 55$. Letting MADN = MAD/.6745, the one-step M-estimator of location is

$$\frac{1.28(\text{MADN})(U - L) + B}{n - L - U}. \tag{8.5}$$

For the data at hand, $1.28(\text{MADN})(U - L) = 16.13$, $n = 6$, so the one-step M-estimator of location is

$$(16.13 + 55)/5 = 14.2.$$

Take another look at the term $B/(n - L - U)$ in Equation 8.5. This is just the average of the values remaining after outliers are eliminated. So our one-step M-estimator almost uses the following method: Begin by identifying outliers using some empirical method that is not subject to masking, as described in Chapter 3. Then remove the outliers and average the values that remain. But for technical reasons, the one-step M-estimator includes a measure of scale as a result of our goal to ensure that our estimator is scale equivariant.

One way to characterize the sensitivity of an estimator to outliers is with the finite sample breakdown point. We have seen that the finite sample breakdown points of the mean, 20 percent trimmed mean, and median are $1/n$, .2, and .5, respectively. Like the median, the one-step M-estimator has a finite sample breakdown point of .5, the highest possible value. But unlike the median, its variance competes very well with the sample mean when sampling from a normal probability curve. Simultaneously, its variance competes very well with the 20 percent trimmed mean when sampling from a heavy-tailed probability curve. In some cases the 20 percent trimmed mean is a bit better, but in other cases the reverse is true. (As for the 10 percent trimmed mean, there are situations where both the 20 percent trimmed mean and an M-estimator with Huber's Ψ are substantially better, but the 10 percent trimmed mean never offers a striking advantage.) So from the point of view of coming up with a single number that estimates a measure of location, the one-step M-estimator looks very appealing.

To elaborate on how the one-step M-estimator described here compares to the 20 percent trimmed mean, we repeat our computer experiment used to create Figure 8.2, but we replace the sample mean with the one-step M-estimator. We generate twenty observations from a normal probability curve, compute the 20 percent trimmed mean and one-step M-estimator, and repeat this process four thousand times. Figure 8.5 shows a plot of the results; as we see, there is little separating the two estimators for this special case. In terms of being close to the population measure of location they are trying to estimate, which in this case is zero, there is little reason to prefer one estimator over the other.

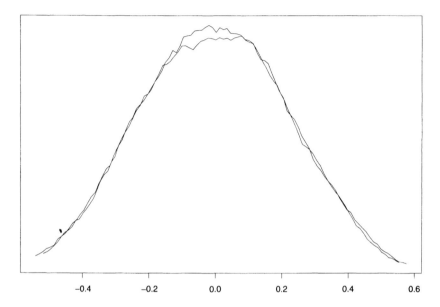

FIGURE 8.5 • A comparison of the 20 percent trimmed mean versus a one-step M-estimator with Huber's Ψ when sampling is from a normal curve. In terms of accuracy, there is little separating the two estimators for this special case.

Next we repeat our computer experiment, but we sample from the mixed normal as was done to create Figure 8.1; the results are shown in Figure 8.6. Again we see that there is little separating the two estimators. So we have an example where sampling is from a heavy-tailed probability curve, yet the 20 percent trimmed mean and one-step M-estimator are very similar in terms of accuracy.

But the M-estimator of location has a higher breakdown point, and this tells us that there are situations where the one-step M-estimator used here will tend to be more accurate than the 20 percent trimmed mean. That is, if there is a sufficient number of outliers, the one-step M-estimator can have a substantially smaller standard error. As an illustration, consider the values

1, 2, 3, 4, 5, 6, 7, 8, 9, 10, 11, 12, 13, 14, 15, 16, 17, 18, 19, 20.

The estimated standard error of the 20 percent trimmed mean and the one-step M-estimator are 1.68 and 1.47, respectively. So there is little separating the two estimators, but the standard errors indicate that one-step M-estimator is a bit more accurate. Now, if we add an outlier by changing the four smallest values to -20 and the four largest values are increased to 40, the estimated standard errors are 1.68 and 2.51. Note that the standard error of the trimmed mean is the same as before because the values that were changed to outliers are ignored in both cases. Now the data indicate that the trimmed mean is more accurate than the one-step M-estimator because its estimated variance is smaller.

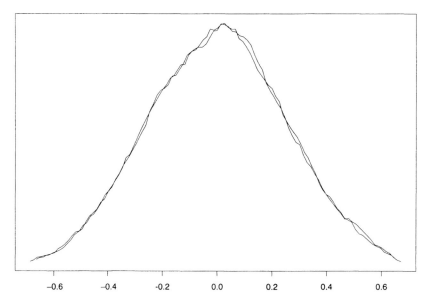

FIGURE 8.6 • A comparison of the 20 percent trimmed mean versus a one-step M-estimator with Huber's Ψ when sampling is from a mixed normal curve. Again, in terms of accuracy, there is little separating the two estimators.

Here is a possible argument for preferring the one-step M-estimator over the 20 percent trimmed mean. Continuing our illustration, notice that with twenty observations, the trimmed mean can handle four outliers among the lower values and four among the higher values. But if instead the lowest five values are decreased to -20, and the largest five are increased to 40, the proportion of outliers in both tails exceeds the breakdown point of the 20 percent trimmed mean. Now the estimated standard errors for the 20 percent trimmed mean and the one-step M-estimator are 8.15 and 4.89, respectively. The trimmed mean has a much higher standard error because the proportion of outliers in both tails exceeds its breakdown point.

But when we look a little closer, we will see that the choice between a 20 percent trimmed mean and an M-estimator with Huber's Ψ is not always straightforward. In fairness to the 20 percent trimmed mean, it is rare to find situations where more than 20 percent trimming is needed to avoid disaster, but in fairness to the one-step M-estimator, such situations may be encountered. When attention is turned to hypothesis testing, the choice between the two approaches depends on what we are trying to accomplish. When comparing populations of individuals based on a measure of location, trimmed means are a bit more appealing for various reasons, but this is not to say that the M-estimator described here has no value. For many purposes there is little separating the two, particularly when sample sizes are not too small. But for small sample sizes, we will see in Chapter 9 that arguments for a 20 percent trimmed mean can be made in some situations. However, when dealing with regression, M-estimators take on a more dominant role.

A SUMMARY OF KEY POINTS

- Two robust estimators of location were introduced: a trimmed mean and an M-estimator.

- The standard errors of the 20 percent trimmed mean and the M-estimator based on Huber's Ψ are only slightly larger than the standard error of the mean when sampling from a normal distribution. But under a very small departure from normality, the standard error of the mean can be substantially higher.

- The 20 percent trimmed mean has a breakdown point of .2. The M-estimator based on Huber's Ψ has a breakdown point of .5. So when sampling from a skewed distribution, their values can be substantially closer to the bulk of the observations than the mean.

BIBLIOGRAPHIC NOTES

It was a paper by Huber (1964) that provided a modern treatment and renewed interest in M-estimators. For theoretical details about M-estimators and trimmed means, see Huber (1981) or Hampel, Ronchetti, Rousseeuw, and Stahel (1986). Staudte and Sheather (1990) summarize mathematical issues at a more intermediate level, but some results are limited to symmetric probability curves. The results reported by Huber and Hampel et al. cover skewed probability curves. These books contain explanations of how an arbitrarily small change in a probability curve can have an arbitrarily large impact on the value of the population mean. For an explanation as to why the biweight and other redescending Ψ's should be avoided, see Freedman and Diaconis (1982). For the earliest paper on how unequal variances affect Student's T, see Box (1954). Apparently the first paper to point out problems with unequal variances when using Student's T and some of its generalizations is Glass, Peckham, and Sanders (1972). For a recent review, see Keselman et al. (1998). For a description of how the standard error of the one-step M-estimator might be estimated and for software to implement the method, see Wilcox (1997).

CHAPTER 9

INFERENCES ABOUT ROBUST MEASURES OF LOCATION

For most purposes, simply estimating a measure of location is not sufficient. There is the issue of assessing the precision of an estimate, and of course there is the related problem of testing hypotheses. How do we test hypotheses or compute confidence intervals with a trimmed mean or an M-estimator of location? To what extent do such methods address the problems with Student's T listed in Chapter 5? These issues are discussed in this chapter.

ESTIMATING THE VARIANCE OF THE TRIMMED MEAN

Perhaps the most obvious and natural approach to making inferences about the population trimmed mean is to use the general strategy developed by Laplace. To do this, the first thing we need is a method for estimating the variance of the sample trimmed mean. That is, we imagine repeating a study infinitely many times, each time computing a trimmed mean based on n observations. The variance of the sample trimmed mean refers to the variation among these infinitely many values, which is consistent with how we view the variance of the mean and median. Our immediate goal is finding some way of estimating this variance based on a single sample of observations.

A cursory consideration of this problem might suggest that it is trivial: Apply the method we used for the sample mean, only now we use the values left after trimming. That is, first compute the sample variance based on the data left after trimming. If there are L such numbers, divide this sample variance by L to estimate VAR($\bar{X}_t$), the variance of the sample trimmed mean. Unfortunately, this approach is highly unsatisfactory, but because it is a common error, some explanation should be given before we proceed.

The first step in understanding what *not* to do when working with a trimmed mean is to take a closer look at how the variance of the sample mean is derived. First consider the situation where we plan to make a single observation, X. As usual, let σ^2 be the population variance associated with X. Now consider the situation where we plan to randomly sample n observations. This means that the n observations are independent. It can be shown that if we sum independent variables, the variance of the sum is the sum of the variances of the individual variables. In symbols, if we let $X_1, \ldots, X_n$ represent our random sample, then

$$\text{VAR}(X_1 + \cdots + X_n) = \text{VAR}(X_1) + \cdots + \text{VAR}(X_n).$$

But the variance of each variable is σ^2, so the right side of this equation is $n\sigma^2$. We obtain the sample mean by dividing the sum of the variables by n, the number of observations. It can be shown that when we divide by n, the variance of the sum is divided by n^2. That is, VAR($\bar{X}$) = $n\sigma^2/n^2 = \sigma^2/n$.

One technical difficulty with the trimmed mean is that, for reasons about to be described, the observations left after trimming are *not* independent— they are dependent. So deriving an expression for the variance of the sample trimmed mean is not straightforward because in general the variance of the sum of dependent variables is not equal to the sum of the individual variances. What is required is some convenient way of taking this dependence into account. But before we describe how this can be done, a more detailed explanation of the dependence among the untrimmed values should be given. This dependence is well known among mathematical statisticians, but it is rarely—if ever—discussed in an applied statistics book. But it is crucial in our quest to understand why certain strategies fail.

To illustrate what is going on, imagine that we plan to randomly sample ten observations from a standard normal curve. Normality is not required here, but it makes the illustration a bit simpler. From basic properties of the normal curve, there is a .84 probability that an observation is greater than -1. Now focus on the first two observations, X_1 and X_2. Independence means that knowing the value of the second variable, X_2, does not alter the probability associated with the first. So if we are told, for example, that the second variable has the value .5, then the probability that the first variable is greater than -1 is still .84.

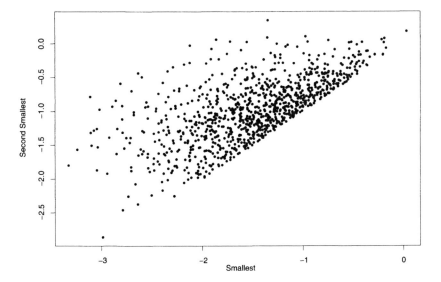

FIGURE 9.1 • A graphical illustration that the two smallest observations among ten randomly sampled observations are dependent. If they were independent, the plot of points would be a random cloud with no visible pattern.

Now suppose that we put our ten observations in order. A common notation for this is $X_{(1)} \leq X_{(2)} \leq \cdots \leq X_{(n)}$. So $X_{(1)}$ represents the smallest value and $X_{(n)}$ the largest. Are these ordered variables independent? The answer is no. To see why, first note that there is some possibility that the smallest observation is greater than -1. That is, the probability is greater than zero. But suppose we are told that the second smallest observation ($X_{(2)}$) has the value -1. Then it is impossible for the smallest value to be greater than -1— that probability is zero. That is, knowing the second smallest value alters the probability associated with the smallest value because the smallest value can never be bigger than the second smallest. (By definition, $X_{(1)}$ is always less than or equal to $X_{(2)}$.) In brief, $X_{(1)}$ and $X_{(2)}$ are dependent. This dependence is illustrated by Figure 9.1, where ten independent observations were generated from a normal curve and the two smallest values are recorded. This process was repeated a thousand times yielding a thousand pairs of observations. The pattern we see in Figure 9.1 indicates that the two smallest observations are dependent. Moreover, this argument generalizes to any pair of the ordered observations, and it generalizes to nonnormal distributions. For example, $X_{(3)}$ and $X_{(4)}$ are dependent, as are $X_{(5)}$ and $X_{(10)}$. Consequently, the method we used to determine an expression for the variance of the sample mean does not readily generalize to the trimmed mean because the trimmed mean is an average of dependent variables. (In fact, there is yet another complication: The variances of the ordered variables are not equal to σ^2.)

A major theoretical advance during the 1960s was the derivation of a mathematical technique that leads to a convenient and practical method for estimating the variance of the sample trimmed mean. The method consists of rewriting the trimmed mean as the average of independent variables so that the strategy used to determine the variance of the sample mean can again be employed. The mathematical details are not given here; readers interested in technical issues can refer to the bibliographic notes at the end of this chapter. Here we first illustrate the resulting method for estimating the variance of the 20 percent trimmed mean and then we try to provide some intuitive explanation. (The method generalizes to any amount of trimming.)

Imagine we observe the values

$$16, 8, 2, 25, 37, 15, 21, 3.$$

The first step is to put the observations in order, yielding

$$2, 3, 8, 15, 16, 21, 25, 37.$$

In our notation for ordering observations, $X_{(1)} = 2$, $X_{(2)} = 3$, $X_{(3)} = 8$, and $X_{(8)} = 37$. With 20 percent trimming, and because we have $n = 8$ observations, the number of observations trimmed from both ends is $g = 1$. (Recall from Chapter 8 that generally, g is $.2n$ rounded down to the nearest integer, assuming 20 percent trimming and that g observations are trimmed.) *Winsorizing* the observations means that rather than drop the g smallest values, we increase their value to the smallest value not trimmed. In our example, the smallest value not trimmed is $X_{(2)} = 3$, so in this case Winsorizing means increasing the smallest value, 2, to 3. More generally, the g smallest values are increased to $X_{(g+1)}$. Simultaneously, the g largest values are decreased to the largest value not trimmed. So in the illustration, 37 is decreased to 25. Using

ESTIMATING THE VARIANCE OF THE TRIMMED MEAN

our more general notation, Winsorizing means that the values $X_{(1)}, \ldots, X_{(g)}$ are increased to $X_{(g+1)}$ and the values $X_{(n-g+1)}, \ldots, X_{(n)}$ are decreased to the value $X_{(n-g)}$. So when we Winsorize our original observations, we now have

$$3, 3, 8, 15, 16, 21, 25, 25.$$

If we had used a 25 percent trimmed mean, then $g = 2$, and the corresponding Winsorized values would be

$$8, 8, 8, 15, 16, 21, 21, 21.$$

That is, the two smallest values would be pulled up to the value 8 and the two largest values pulled down to 21.

The next step is to compute the sample variance of the Winsorized values. The resulting value is called the *Winsorized sample variance*. For example, to compute the 20 percent Winsorized sample variance for the values 2, 3, 8, 15, 16, 21, 25, and 37, first Winsorize them, yielding 3, 3, 8, 15, 16, 21, 25, 25 and then compute the sample variance using these Winsorized values. The mean of the Winsorized values is called the *Winsorized sample mean* and is equal to

$$\bar{X}_w = \frac{1}{8}(3 + 3 + 8 + 15 + 16 + 21 + 25 + 25) = 14.5.$$

The next step when computing the Winsorized sample variance is to subtract the Winsorized mean from each of the Winsorized values, square each result, and then sum. Finally, divide by $n - 1$, the number of observations minus 1, as when computing the sample variance, s^2. In our example, $n = 8$, so the Winsorized sample variance is

$$s_w^2 = \frac{1}{7}[(3 - 14.5)^2 + (3 - 14.5)^2 + (8 - 14.5)^2 + \cdots + (25 - 14.5)^2] = 81.7.$$

In a more general notation, if we let $W_1, \ldots, W_n$ represent the Winsorized values corresponding to $X_1, \ldots, X_n$, then the Winsorized sample variance is

$$s_w^2 = \frac{1}{n-1}[(W_1 - \bar{X}_w)^2 + \cdots + (W_n - \bar{X}_w)^2],$$

where

$$\bar{X}_w = \frac{1}{n}(W_1 + \cdots + W_n)$$

is the Winsorized sample mean.

Finally, we can estimate $\text{VAR}(\bar{X}_t)$, the variance of the trimmed mean. With 20 percent trimming, the estimate is

$$\frac{s_w^2}{.36n}, \tag{9.1}$$

the Winsorized sample variance divided by .36 times the sample size. Continuing our illustration, the estimated variance of the 20 percent trimmed mean is $81.7/(.36(8)) = 28.4$. That is, if we were to repeatedly sample eight observations, each time computing the 20 percent trimmed mean, we would estimate that the variance among the resulting trimmed means is 28.4. So the estimated standard error of the sample trimmed mean (the square root of the variance of the trimmed mean) is $\sqrt{28.4} = 5.3$. For the more general case where the amount of trimming is γ ($0 \leq \gamma < .5$), the estimated variance of the trimmed mean is

$$\frac{s_w^2}{(1-2\gamma)^2 n}. \tag{9.2}$$

As is probably evident, this method for estimating the variance of the sample trimmed mean is not intuitive. There is no obvious reason why we should Winsorize, and we must divide by $(1-\gamma)^2$, a result that follows from a mathematical analysis that goes beyond the scope of this book. But this method is not only indicated by theoretical results; it has great practical value as well.

Although complete technical details are not given here, perhaps some indication as to why we divide by $(1-2\gamma)^2$ can be given. For illustrative purposes, again imagine that we intend to use 20 percent trimming and that sampling is from a normal curve. Then 20 percent trimming is tantamount to chopping off the two tails of the normal curve, leaving what is shown in Figure 9.2. Note that the area under the curve left after trimming is .6. That is, the area under the normal curve left after trimming the tails is no longer 1, so from a technical point of view we no longer have a probability curve. (By definition, the area under any probability curve must be 1.) To transform the curve in Figure 9.2 so that it is a probability curve, we must divide by .6. More generally, when trimming the amount γ from each tail, we must divide by $1 - 2\gamma$. It is stressed that from a mathematical point of view, this explanation is not remotely satisfactory—it is merely suggestive. And no explanation has been given as to why we must also Winsorize when deriving a method for estimating the variance of the sample trimmed mean.

The expression for the estimated variance of a trimmed mean gives us a formal way of indicating why the 20 percent trimmed mean is a more accurate

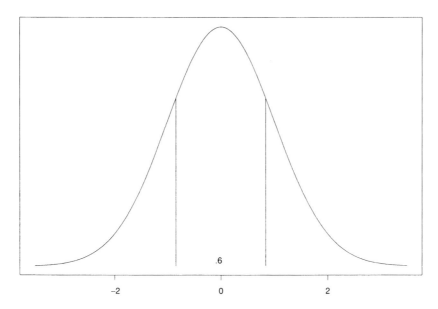

FIGURE 9.2 • Trimming a normal curve by 20 percent means that we focus on the middle 60 percent of the curve. That is, we discard the two tails. But eliminating the tails means that the area under the curve is no longer 1, as required by a probability curve.

estimator than the mean when sampling from a mixed normal, as was illustrated in Figure 8.1 of Chapter 8. Data indicate that the 20 percent trimmed mean will be more accurate when it has a smaller estimated variance, meaning that $s_w^2/.36 < s^2$. Notice that the finite sample breakdown point of the 20 percent Winsorized variance is .2. So it takes more than 20 percent of the observations to be outliers for s_w^2 to be inflated by unusually large or small values. Consequently, the variance of the trimmed mean is relatively unaffected when we sample, for example, from the mixed normal described in Chapter 7, a situation where outliers are common, compared to sampling from a normal curve. In general, we would expect the Winsorized variance to be smaller than the sample variance s^2 because the Winsorized variance pulls in extreme values that tend to inflate s^2. Of course, this is true even when sampling from a skewed distribution. However, this is countered by the fact that with 20 percent trimming, the variance of the trimmed mean is based not just on s_w^2, but by s_w^2 divided by .36. This makes it possible for the sample mean to have a smaller variance, such as when sampling from a normal curve, as was illustrated in Figure 8.2. But typically the improvement using the mean is small, and Figure 8.1 illustrated that for a very small departure from normality, the variance of the trimmed mean can be substantially smaller than the mean.

INFERENCES ABOUT THE POPULATION TRIMMED MEAN

Chapter 3 described the central limit theorem for the sample mean, which says that the probability curve associated with the sample mean approaches a normal curve as the sample size gets large. The method for deriving the estimate of the variance of the trimmed mean can be used to describe general conditions under which the sample trimmed mean also has a normal distribution. That is, like the sample mean, if we were to repeat an experiment infinitely many times, each time computing the trimmed mean, the plot of the trimmed means would be approximately normal, provided each trimmed mean is based on a reasonably large sample size. Moreover, the sample trimmed means will be centered around the population trimmed mean. As a result, we can use Laplace's method to compute a .95 confidence interval for the population trimmed mean. With 20 percent trimming it is

$$\left(\bar{X}_t - 1.96 \frac{s_w}{.6\sqrt{n}},\ \bar{X}_t + 1.96 \frac{s_w}{.6\sqrt{n}}\right). \tag{9.3}$$

If we set the amount of trimming to zero, Equation 9.3 reduces to Laplace's confidence interval for the mean.

We saw that when working with the mean, we could improve Laplace's method when the sample sizes are small by using Student's T distribution. The method is based in part on determining the probability curve for T when sampling from a normal curve. An analog of this method can be used for a trimmed mean. In general, inferences about the population 20 percent trimmed mean (μ_t) can be made if we can get a reasonably accurate approximation of the probability curve associated with

$$T_t = \frac{\bar{X}_t - \mu_t}{s_w/(.6\sqrt{n})}. \tag{9.4}$$

It turns out that a reasonable approximation of the probability curve for T_t is with the probability curve for T when using means and the sample size is $n - 2g$. In more conventional terms, T_t will have, approximately, a Student's T distribution with $n - 2g - 1$ degrees of freedom. So standard tables of

Student's T can be used to test hypotheses. For example, if we have $n = 12$ observations, then with 20 percent trimming, $g = 2$, and the degrees of freedom are $12 - 4 - 1 = 7$. From tables of Student's T distribution, we see that with probability .95 and seven degrees of freedom, T will have a value between -2.365 and 2.365. That is, $P(-2.365 \leq T \leq 2.365) = .95$. This means that when we compute a 20 percent trimmed mean based on twelve observations (in which case there are again seven degrees of freedom), then it will be approximately true that T_t will be between -2.365 and 2.365 with probability .95. That is, $P(-2.365 \leq T_t \leq 2.365) = .95$. To test the hypothesis $H_0 : \mu_t = \mu_0$, where again μ_0 is some specified constant, we compute T_t with Equation 9.4, but with μ_t replaced by μ_0. If we reject when $|T_t| > 2.365$, the probability of a Type I error will be approximately .05. Or, an approximate .95 confidence interval for the population 20 percent trimmed mean is

$$\left(\bar{X}_t - 2.365 \frac{s_w}{.6\sqrt{n}}, \; \bar{X}_t + 2.365 \frac{s_w}{.6\sqrt{n}} \right),$$

and we reject $H_0 : \mu_t = \mu_0$ if this interval does not contain μ_0.

The method just described for making inferences about the population trimmed mean is fairly accurate when sampling from a normal curve. But we saw in Chapter 5 that nonnormality can result in highly inaccurate inferences about the mean when using T, so there is the issue of how nonnormality affects T_t. Theoretical results, and studies on how T_t performs when sample sizes are small (simulation studies) indicate that the more we trim, the smaller the effect of nonnormality in terms of Type I errors, bias, and probability coverage. In some situations, problems decrease precipitously as the amount of trimming increases, up to about 20 percent. With higher amounts of trimming, performance continues to improve, but at a rather reduced rate. Another advantage associated with the 20 percent trimmed mean is that problems disappear more quickly than when using the mean, as the sample size gets large. In fact, in terms of Type I error probabilities and probability coverage, using a 20 percent trimmed mean can be substantially more accurate than any method based on means, including the percentile t bootstrap method covered in Chapter 6. Unfortunately, however, we still find situations where control over the probability of a Type I error and probability coverage are deemed unsatisfactory. That is, trimming reduces practical problems, in some cases substantially, but it does not eliminate them. For example, we saw a situation in Chapter 5 where, when using Student's T with means, we needed about two hundred observations to control the probability of a Type I error. If we switch to a 20 percent trimmed mean, we get good control over the probability of a Type I error with one hundred observations, but problems persist with $n = 20$.

So again we have made progress, but more needs to be done. Because theory tells us that the percentile t bootstrap improves control over the probability of a Type I error when using the mean and that trimming improves matters, a natural strategy is to combine the two methods. When we do, we get even better control over the probability of a Type I error. In fact, all indications are that we can avoid Type I error probabilities substantially higher than the nominal .05 level with sample sizes as small as twelve (Wilcox, 1996b).

To apply the percentile t bootstrap method to the problem at hand, we again generate a bootstrap sample on a computer, but now we compute

$$T_t^* = \frac{\bar{X}_t^* - \bar{X}_t}{s_w^*/(.6\sqrt{n})}, \tag{9.5}$$

where $\bar{X}_t^*$ and s_w^* are the 20 percent trimmed mean and Winsorized standard deviation based on the bootstrap sample. Next, repeat this process B times yielding B values for T_t^*. When using the percentile t method in conjunction with a 20 percent trimmed mean, all indications are that $B = 599$ is a good choice when computing a .95 confidence interval, so the bootstrap T_t^* values can be labeled $T_{t1}^*, \ldots, T_{t599}^*$. (On rare occasions, there is a practical advantage to using $B = 599$ rather than $B = 600$; no situations have been found where the reverse is true, so $B = 599$ is recommended for general use.) Next, using the bootstrap values just generated, determine the values t_L^* and t_U^* such that the middle 95 percent of the bootstrap values are between these two numbers. With $B = 599$, these two numbers are $T_{t(15)}^*$ and $T_{t(584)}^*$, where as usual $T_{t(1)}^* \leq \cdots \leq T_{t(599)}^*$ represent the five hundred ninety-nine bootstrap values written in ascending order. Then an approximate .95 confidence interval for the population trimmed mean is

$$\left(\bar{X}_t - T_{(U)}^* \frac{s_w}{.6\sqrt{n}}, \bar{X}_t - T_{(L)}^* \frac{s_w}{.6\sqrt{n}} \right). \tag{9.6}$$

So, as was the case in Chapter 6 when working with the mean, we use the bootstrap to estimate the probability curve of T_t, which gives us an estimate of where we will find the middle 95 percent of the T_t values if we were to repeat our experiment many times, and the result is a .95 confidence interval that is relatively accurate. Indeed, in terms of getting accurate probability coverage or controlling the probability of a Type I error over a fairly broad range of situations, this is the method to beat.

THE RELATIVE MERITS OF USING A TRIMMED MEAN VERSUS A MEAN

The combination of the percentile t bootstrap and the 20 percent trimmed mean, as just described, addresses all the problems with Student's T test listed in Chapter 5, a result supported by both theory and simulation studies. The problem of *bias* (power going down as we move away from the null hypothesis) appears to be negligible, we get vastly more accurate confidence intervals in situations where all methods based on means are deemed unsatisfactory, we get better control over Type I error probabilities, we get better power in a wider range of situations, and in the event sampling is from a normal curve, using means offers only a slight advantage. If sampling is from a probability curve that is symmetric, the population mean and trimmed mean are identical. But if sampling is from a skewed curve, as shown in Figure 8.3, a 20 percent trimmed mean is closer to the most likely values and provides a better reflection of the typical individual under study.

Chapter 5 illustrated yet another peculiarity of Student's T. Consider any situation where the goal is to test $H_0 : \mu = \mu_0$, where μ_0 is some specified constant. Further imagine we get a sample mean greater than μ_0 and that H_0 is rejected. If the largest observed value is increased, of course the sample mean increases. This might suggest that we have more compelling evidence to reject our null hypothesis, but if the largest value is increased sufficiently, we will no longer reject. The reason is that the sample variance increases, more rapidly than the sample mean in the sense that the value of T (given by Equation 5.1) decreases, which means we are no longer able to reject. Put another way, as we increase the largest value, the .95 confidence interval is centered around a sample mean that is increasing, but this is more than offset by an increasingly wider confidence interval due to the increasing magnitude of the standard error of the sample mean.

Note, however, that virtually nothing happens to the confidence interval for the trimmed mean because as we increase the largest observation, the sample Winsorized variance is not altered. For example, if we compute the Winsorized variance for the values

$$1\ 2\ 3\ 4\ 5\ 6\ 7\ 8\ 9\ 10\ 11\ 12\ 13\ 14\ 15,$$

we get $s_w^2 = 11.1$. If we increase the largest value from 15 to 400, again the Winsorized variance is 11.1. Consequently, the confidence interval for the trimmed mean remains relatively unaffected. (When using the percentile t method, it might be altered very slightly due to the nature of the bootstrap technique.) We need to increase more than 20 percent of the largest values to have an impact on the Winsorized variance.

THE TWO-SAMPLE CASE

The bootstrap method for computing a confidence for the population trimmed mean is readily extended to the situation where we want to compare two independent groups of individuals, and it can be used to test $H_0 : \mu_{t1} = \mu_{t2}$, the hypothesis that the groups have identical population trimmed means. To begin, compute the trimmed mean and Winsorized variance for each group and label the trimmed means $\bar{X}_{t1}$ and $\bar{X}_{t2}$ and the Winsorized variances s_{w1}^2 and s_{w2}^2. Again we need to estimate the variance associated with the sample trimmed means. Currently, however, it seems that there is merit to using a slightly different approach from the method previously described. As suggested by K. Yuen, we use

$$d_1 = \frac{(n_1 - 1)s_{w1}^2}{h_1(h_1 - 1)}$$

to estimate the variance of the sample trimmed mean $\bar{X}_{t1}$, rather than use $s_{w1}^2/(.36n_1)$, where h_1 is the number of observations left after trimming. The two methods give very similar results, but in terms of Type I error probabilities, Yuen's method has been found to perform slightly better when sample sizes are small. In a similar manner, the variance associated with the second trimmed mean, $\bar{X}_{t2}$, is estimated with

$$d_2 = \frac{(n_2 - 1)s_{w2}^2}{h_2(h_2 - 1)}.$$

Analogous to the one-sample case, inferences about the difference between the population trimmed means can be made if we can approximate the probability curve associated with

$$W = \frac{(\bar{X}_{t1} - \bar{X}_{t2}) - (\mu_{t1} - \mu_{t2})}{\sqrt{d_1 + d_2}}. \tag{9.7}$$

A bootstrap approximation of this probability curve is obtained in a manner similar to the one-sample case. First, generate bootstrap samples from each group and compute

$$W^* = \frac{(\bar{X}_{t1}^* - \bar{X}_{t2}^*) - (\bar{X}_{t1} - \bar{X}_{t2})}{\sqrt{d_1^* + d_2^*}}, \tag{9.8}$$

where d_1^* and d_2^* are the values of d_1 and d_2 based on the bootstrap samples, and of course $\bar{X}_{t1}^*$ and $\bar{X}_{t2}^*$ are the bootstrap trimmed means. Then repeat this B times yielding B bootstrap values for W, which we label $W_1^*, \ldots, W_B^*$. These bootstrap values can be used to compute a confidence interval for the difference between the population means using what is essentially the same strategy as in the one-sample case. The choice $B = 599$ has been found to perform relatively well with 20 percent trimming when the goal is to compute a .95 confidence interval. In formal terms, order the bootstrap values, yielding $W_{(1)}^* \leq \cdots \leq W_{(B)}^*$. With $B = 599$, set $L = 15$ and $U = 584$. Then the middle 95 percent of the W^* values lie between $W_{(L)}^*$ and $W_{(U)}^*$. So an approximate .95 confidence interval for the difference between the population trimmed means is

$$[(\bar{X}_{t1} - \bar{X}_{t2}) - W_{(U)}^*\sqrt{d_1 + d_2}, (\bar{X}_{t1} - \bar{X}_{t2}) - W_{(L)}^*\sqrt{d_1 + d_2}]. \quad (9.9)$$

If this interval does not contain zero, then reject the hypothesis that the population trimmed means are equal. That is, reject $H_0 : \mu_{t1} = \mu_{t2}$.

As an illustration, we compare the trimmed means for two groups of subjects from a study on self-awareness. The data were collected by E. Dana and are

Group 1: 77 87 87 114 151 210 219 246 253 262 296 299 306 376 428 515 666 1310 2611

Group 2: 59 106 174 207 219 237 313 365 458 497 515 529 557 615 625 645 973 1065 3215.

The sample trimmed means (with 20 percent trimming) are 282.7 and 444.77. Applying the percentile t bootstrap method, the .95 confidence interval for the difference between the trimmed means is $(-340.6, -12.1)$. This interval does not contain 0, so you would reject the hypothesis of equal population trimmed means. That is, the data suggest that the typical subject in the first group tends to score lower than the typical subject in the second. In contrast, comparing means with no bootstrap (using what is called *Welch's test*, a method generally more accurate than Student's T), the .95 confidence interval is $(-574.1, 273.0)$. In this case you would not reject, and if we used Student's T despite its technical problems, we get a similar result and again would not reject. That is, we would no longer come to the conclusion that the typical subject in the first group differs from the typical subject in the second.

With sufficiently large sample sizes, the bootstrap can be avoided and inferences can be made by approximating the probability curve associated with W using Student's T distribution (with degrees of freedom estimated from the data.) This was the approach originally suggested by Yuen. There is uncertainty as to how large the sample sizes must be, but an educated guess is that if both sample sizes are at least one hundred, Yuen's method can be used. There is reason to hope that even smaller sample sizes might justify replacing the bootstrap with Yuen's procedure, but this remains to be seen.

POWER USING TRIMMED MEANS VERSUS MEANS

In the last example, notice that the confidence interval for the difference between the trimmed means is considerably shorter than the confidence interval for the difference between the means. The reason is that there are outliers among the observations that greatly reduce power when comparing the means, but by switching to trimmed means, the loss of power is greatly reduced. A crude explanation is that the trimmed means ignore the outliers, so it's not surprising that we get more power than when using means. A more precise explanation is that the standard error of the sample trimmed mean is less than the standard error of the mean because the standard error of the trimmed mean is based in part on the Winsorized standard deviation.

To drive home a potential advantage to using a trimmed mean, look at the left panel of Figure 9.3, which shows two normal probability curves. When sampling twenty-five observations from each curve and testing the hypothesis of equal means, power is approximately .96 using Student's T and .93 using Welch's test. That is, we have a 96 percent chance of correctly detecting a difference between these two curves when using Student's T. In contrast, using trimmed means, power is .89, a bit less because nothing beats means when sampling from a normal curve. Now imagine that we sample from the two curves shown in the right panel of Figure 9.3. As is evident, there is little visible difference compared to the left panel. Yet when comparing means, power is only .278 with Student's T versus .784 when using a 20 percent trimmed mean. In fact, even smaller departures from normality can result in very low power when using means than when using a 20 percent trimmed mean.

A criticism of the illustration in Figure 9.3 is the following. For skewed distributions, the population mean and trimmed mean differ. An implication is that for two groups of individuals, the difference between the population

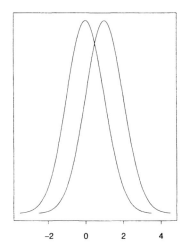 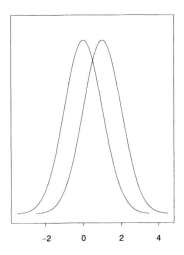

FIGURE 9.3 • For the normal curves shown in the left panel, power is slightly higher when comparing means rather than 20 percent trimmed means. But for the right panel, power is substantially higher using 20 percent trimmed means. This illustrates that slight departures from normality can drastically reduce power when using means versus more modern methods.

means might be larger than the difference between the population trimmed means. If this is the case, this might result in more power when using a method based on the mean. Again, however, this must be weighed against the variance of the trimmed mean versus the variance of the mean. Even if the difference between the population means is larger, the variance of the sample means might exceed the variance of the trimmed means to the point that we are still more likely to detect a true difference using trimmed means. As already noted, skewness can result in a biased test (power can decrease as the difference between the means increases), and switching to trimmed means can reduce this problem substantially. The only certainty is that neither approach is always optimal. However, if any value can be given to experience, it is common to find situations where one fails to find a difference with means, but a difference is found when using trimmed means. It is rare for the reverse to happen, but it does happen.

INFERENCES ABOUT M-ESTIMATORS

What about the M-estimator (based on Huber's Ψ) described in Chapter 8? How do we estimate its variance and how should we compute a confidence interval? An expression for the variance of an M-estimator has been derived and can be estimated based on observations we make. But the resulting method is rather complicated and does not yield confidence intervals that are satisfactory when using an analog of Laplace's method. When sampling from a perfectly symmetric probability curve, adjustments can be made, based on the sample size, that give fairly accurate results. (The adjustment is similar in spirit to how Student improved upon Laplace's method when working with means.) But under even slight departures from a perfectly symmetric curve, the method breaks down and cannot be recommended, at least when the number of observations is small or moderately large. It is unknown how large the sample size must be before practical problems are eliminated.

Because the percentile t method performs so well when using a 20 percent trimmed mean, a natural guess is to use this approach with an M-estimator, but this is not quite satisfactory when sample sizes are small. What performs better is the percentile bootstrap. This is somewhat surprising because the percentile method performs poorly with means compared to using the percentile t.

For convenience, let $\hat{\theta}$ represent the value of the one-step M-estimator given by Equation 8.5 of Chapter 8. To compute a .95 confidence interval for the population value of the M-estimator, begin by generating B bootstrap estimates:

$$\hat{\theta}_1^*, \ldots, \hat{\theta}_B^*.$$

That is, repeatedly generate bootstrap samples and compute the one-step M-estimator given by Equation 8.5. A .95 confidence interval for the population analog of the M-estimator is given by the middle 95 percent of the bootstrap values. In essence, we simply apply the percentile method for means, as described in Chapter 6, but we compute bootstrap M-estimators rather than sample means. With a sample size of at least twenty, all indications are that an accurate .95 confidence interval can be computed with $B = 399$. For smaller sample sizes, the probability coverage can be unsatisfactory, and increasing B to 599 does not correct this problem. So, for very small samples sizes, if our criterion is accurate probability coverage, the percentile t bootstrap with a 20 percent trimmed mean is more satisfactory. But with at least twenty observations, the method just described performs about as well.

THE TWO-SAMPLE CASE USING AN M-ESTIMATOR

The percentile method is easily extended to the situation where the goal is to compare two (independent) groups of individuals using M-estimators of location. You merely generate B bootstrap samples from the first group, each time computing the M-estimator of location, yielding

$$\hat{\theta}_{11}^*, \ldots, \hat{\theta}_{1B}^*.$$

Repeat this for the second group, yielding

$$\hat{\theta}_{21}^*, \ldots, \hat{\theta}_{2B}^*,$$

and then form the differences

$$D_1^* = \hat{\theta}_{11}^* - \hat{\theta}_{21}^*, \ldots, D_B^* = \hat{\theta}_{1B}^* - \hat{\theta}_{2B}^*.$$

The middle 95 percent of these difference values yields an approximate 95 percent confidence interval. All indications are that $B = 399$ gives reasonably good results and that there is little or no advantage to using a larger number of bootstrap values. Again, however, when sample sizes are small, more accurate probability coverage might be obtained with 20 percent trimmed means.

SOME REMAINING ISSUES

The percentile t bootstrap method, used with 20 percent trimmed means, does a remarkable job of reducing practical problems associated with making inferences about means with Student's T. In fact, even with sample sizes as small as eleven, Type I error probabilities substantially larger than the nominal level of $\alpha = .05$ can be avoided. There are, however, at least three concerns that remain. The first is whether an alternative method can be found that performs about as well as the percentile t method but gives shorter confidence intervals. Recent results indicate that the answer is yes (Wilcox, in press b). The method consists of simply switching to the percentile method, again using 20 percent trimming. This is somewhat unexpected because the percentile method performs poorly with means relative to the percentile t. But it appears that when working with estimators with a reasonably high finite sample breakdown point, the reverse is true. When comparing two groups only, very recent results indicate that the percentile bootstrap improves upon the percentile t

bootstrap method when using 20 percent trimmed means. And when comparing more than two groups, it appears that the advantages of the percentile bootstrap improve. In addition, there are theoretical results that suggest modifying the bootstrap method: Winsorize before drawing bootstrap samples. This has the potential of providing even shorter confidence intervals, but the issue of how much to Winsorize remains an important concern. Current results suggest using an amount that is less than the amount of trimming.

A second issue has to do with what is called an *equal-tailed test*. As previously indicated, when we test the hypothesis of equal means with Student's T, under normality the probability curve associated with T is symmetric about zero when the hypothesis is true. Moreover, we reject if the value of T is sufficiently large or small. The method is designed so that when testing at the .05 level and the null hypothesis is true, the probability of rejecting because T is large is .025, and the probability is again .025 due to T being small, in which case the probability of a Type I error is $.025 + .025 = .05$. This is an example of an equal-tailed test, meaning that the probability of rejecting due to a large T value is the same as the probability of rejecting due to T being small. But because of nonnormality, the probability curve for T may not be symmetric about zero. If it is asymmetric, the probability of rejecting due to a large T value might be .01, and the corresponding probability for a small T value is .04, in which case the probability of a Type I error is $.04 + .01 = .05$, the same as before, but it is not equal tailed. In fact, as already indicated, the tail probabilities can exceed the nominal level by a considerable amount.

One reason this is a concern is illustrated by the following problem. In the pharmaceutical industry, regulatory agencies allow a generic drug to be marketed if its manufacturer can demonstrate that the generic product is equivalent to the brand-name product. Often a portion of the process of establishing equivalence has to do with determining whether the population mean of an appropriate measure for the generic drug is close to the mean for the brand-name drug. Based on constants μ_L and μ_U specified by regulatory agencies, a necessary condition for two drugs to be declared equivalent is that the difference between the means has a value between μ_L and μ_U. For example, if $\mu_L = -2$ and $\mu_U = 2$, a necessary condition for the drugs to be declared equivalent is that the difference between the population means $(\mu_1 - \mu_2)$ is somewhere between -2 and 2. An approach to establishing equivalence is to test

$$H_0 : \mu_1 - \mu_2 \leq \mu_L \text{ or } \mu_1 - \mu_2 \geq \mu_U$$

versus

$$H_1 : \mu_L < \mu_1 - \mu_2 < \mu_U.$$

So if we reject H_0, the conclusion is that the population means do not differ very much. (The difference between the means lies between μ_L and μ_U.)

A natural approach to testing H_0 is to perform two one-sided tests, namely,

$$H_{01} : \mu_1 - \mu_2 \leq \mu_L$$

and

$$H_{02} : \mu_1 - \mu_2 \geq \mu_U.$$

Without going into detail, it turns out that the probability of a Type I error associated with H_0 depends on the maximum of the Type I error probabilities associated with H_{01} and H_{02}. If the goal is to have a Type I error probability equal to .025, this will be achieved if the probability of a Type I error associated with both H_{01} and H_{02} is .025. But if one of these tests has probability .04 and the other has probability .01, it can be shown that the probability of a Type I error when testing H_0 is .04, the larger of the two probabilities. So in effect, problems due to nonnormality are exacerbated. There is some indication that if we compare 20 percent trimmed means with a percentile bootstrap, we are better able to ensure an equal-tailed test than when using a percentile t bootstrap.

Yet another lingering concern about the percentile t method is that when comparing multiple groups of individuals with a 20 percent trimmed mean, the probability of a Type I error can drop well below the nominal level. Generally, this problem does not arise, but it would be nice if it could be avoided altogether. In practical terms, there are situations where power could be higher because one is testing hypotheses at the .015 level versus what was intended, which is testing at the .05 level. This implies that power will be lower than with a method where the actual probability of a Type I error is .05 because, as noted in Chapter 5, power is related to the probability of a Type I error. Again there is evidence that switching to a percentile bootstrap method, still using 20 percent trimmed means, provides an effective way of addressing this problem (Wilcox, in press b).

A SUMMARY OF KEY POINTS

- A seemingly natural but incorrect method for estimating the squared standard error of a trimmed mean is to apply the method for the sample mean to the observations left after trimming. A technically correct estimate is obtained via Equation 9.2, which is based in part on the Winsorized variance. In some cases the correct estimate is substantially smaller than the incorrect estimate. So when testing hypotheses, the correct estimate can mean more power.

- Given an estimate of the standard error of a trimmed mean, confidence intervals can be computed and hypotheses can be tested. Theory and simulations indicate that practical problems associated with Student's T for means are reduced substantially, but not all problems are eliminated. The percentile t bootstrap reduces these problems even further.

- Equation 9.4 indicates how to compare the trimmed means of two independent groups. Under normality, little power is lost using trimmed means rather than means. But for a very small departure from normality, using a 20 percent trimmed mean can result in substantially more power.

- Confidence intervals can be computed and hypotheses tested using an M-estimator in conjunction with a percentile bootstrap technique. Again the method competes well with conventional methods for means under normality, but substantially better power is possible when sampling from a nonnormal distribution instead. With large sample sizes it is possible to avoid the bootstrap when using an M-estimator, but it is unknown how large the sample size must be.

BIBLIOGRAPHIC NOTES

For an excellent discussion of how to establish equivalence, see Berger and Hsu (1996). Expressions for the variance of a trimmed mean or an M-estimator of location can be derived using what is called the *influence function*. Details can be found in Huber (1981) and Hampel, Ronchetti, Rousseeuw, and Stahel (1986). Staudte and Sheather (1990) also summarize this method, but the results in Huber (1981) are more general—Huber's results cover skewed probability curves. For the first paper on how to compare trimmed means, see Yuen (1974). For more recent results and results on comparing M-estimators, see Wilcox (1997). Luh and Guo (1999) suggest yet another method for comparing trimmed means that avoids the bootstrap. Generally the method seems to perform well, but for skewed distributions it might be unsatisfactory unless the occurrence of outliers is sufficiently common or the sample sizes are sufficiently large.

CHAPTER *10*

MEASURES OF ASSOCIATION

Pearson's correlation coefficient ρ, introduced in Chapter 6, is ubiquitous in applied research. It is employed even more than the least squares estimate of the slope of a regression line and is often the lone tool used to detect and describe an association between two variables. As is evident from results covered in previous chapters, the estimate of ρ, r can miss important associations. Even if the value of ρ could be determined exactly, its value can be relatively unrevealing. The goal in this chapter is to introduce some new tools for detecting and describing the association between variables that deal with some of the practical problems associated with r. But first we look more closely at how r and ρ are interpreted and the features of data that influence their values.

WHAT DOES PEARSON'S CORRELATION TELL US?

As previously indicated, ρ is exactly equal to zero when two measures, say X and Y, are independent. Moreover, it can be shown that both ρ and its estimate r always have values between -1 and 1. If all points lie on a straight line with a positive slope $r = 1$, and if the slope is negative, $r = -1$. Pearson's correlation is related to the slope of the least squares regression line (β_1) in the following manner:

$$\beta_1 = \rho \frac{\sigma_y}{\sigma_x}. \tag{10.1}$$

So the slope of the least squares regression line is determined by three quantities: the correlation, the variance associated with the X values, and the variance associated with the Y values. As is evident, if $\rho > 0$, the slope is positive, and the reverse is true if ρ is negative. Consistent with Equation 10.1 the least squares estimate of the slope can be written as

$$b_1 = r \frac{s_y}{s_x}, \tag{10.2}$$

a result that will prove useful in Chapter 11. So when we estimate ρ with r, the sign of r tells us whether the least squares estimate of the regression line will be positive or negative, but it does not tell us how quickly Y changes with X except in a special case to be described.

As noted in Chapter 6, r is not resistant to outliers, so for the bulk of the observations it might poorly reflect whether there is a positive or negative association between some outcome measure (Y) and some predictor (X). Said another way, r tells us whether the slope of the least squares regression line will be positive or negative, but we have already seen that the least squares regression line might poorly reflect how the bulk of the observations are associated. Figure 10.1 illustrates that even one unusual point can mask an association. As can be seen, all of the points lie on a perfectly straight line except one, yet $r = 0$. But there are other features of data that influence the value of r that should be described so we can better decipher what r is telling us.

To illustrate one of these features, first consider a situation where points are centered around the regression line $Y = X$. If we observe the points shown in Figure 10.2, Pearson's correlation is .92. Now suppose that the points are centered around the same line, but they are further from the line, as shown in the right panel of Figure 10.2. In terms of regression, the residuals are larger in the right panel than in the left, meaning that there is more uncertainty about the mean of Y given X. Now the correlation is .42. But look at Figure 10.3. These are the same points shown in left panel of Figure 10.2, but they have been rotated so that the slope of the line around which they are centered has

What Does Pearson's Correlation Tell Us?

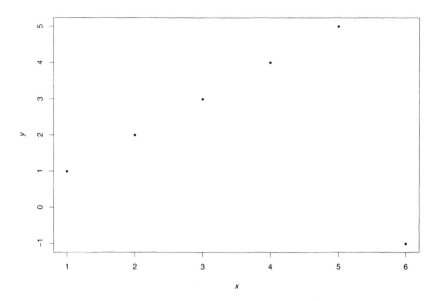

FIGURE 10.1 • Even if all but one point lie on a straight line, the one unusual point can greatly affect r.

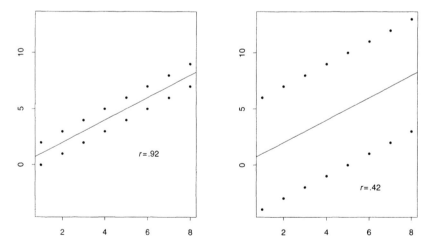

FIGURE 10.2 • The left panel shows some points centered around a straight line having a slope of one. As indicated, $r = .92$. If we increase the distance of the points from the regression line, as shown in the right panel, r drops to .42. That is, the magnitude of the residuals are related to r.

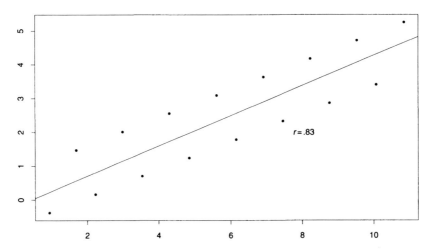

FIGURE 10.3 • The same points as in the left panel of Figure 10.2 rotated so that they are centered around a line with a slope of .44. Rotating the points lowered the correlation from .92 to .83.

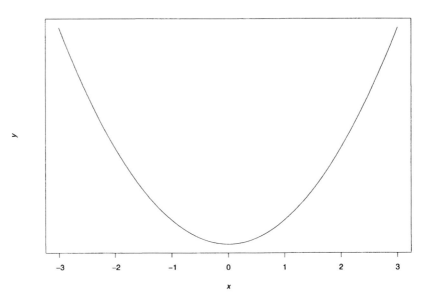

FIGURE 10.4 • Generally, curvature influences the magnitude of r. Here there is an exact assocation, but the correlation is zero.

been decreased from 1 to .44. Now the correlation is .83, so we see that even when there is a linear association between X and Y, both the slope of the line around which the points are clustered and the magnitude of the residuals are reflected by r. If we rotate the points until they are centered around the x-axis, $r = 0$.

Yet another factor influencing the value of r is the degree to which there is a nonlinear association. In Figure 10.4 there is an exact association between X and Y, yet the correlation is zero.

Many books point out that restricting the range of X can lower $|r|$, the amount r differs from zero. Figure 10.1 illustrates that restricting the range of X can increase $|r|$; the data in Figure 7.10 provide yet another illustration that restricting the range of X might substantially increase $|r|$. Pearson's correlation is $-.03$, but if we eliminate the right-most six points by excluding points with $X > 100$, $r = -.39$.

In summary, the following factors influence the value of r: (1) the magnitude of the residuals, (2) the slope of the line around which points are clustered, (3) outliers (as was illustrated by Figure 6.6 in Section 6.6), (4) curvature, and (5) a restriction of range. So if we are told r, and nothing else, we cannot deduce much about the details of how X and Y are related. The good news is that there is a collection of methods that help us get a better picture and understanding about the association between X and Y, some of which are described here.

A COMMENT ON CURVATURE

Before continuing, a comment on curvilinear relationships should be made. It is quite common to think of curvature in terms of a quadratic relationship. In Figure 10.4, $Y = X^2$, and a seemingly natural strategy for modeling curvature would be to use a regression line with the form $\hat{Y} = \beta_1 X + \beta_2 X^2 + \beta_0$. Other strategies include replacing X^2 with $\sqrt{X}$, or $\log(X)$, or $1/X$. However, recently developed methods for studying curvature suggest that these approaches can often be rather unsatisfactory. For example, it is quite common for there to be a reasonably linear association between X and Y over some range of X values, but outside this range the association might change substantially or disappear altogether. Figure 10.5 illustrates this point using data introduced in Chapter 4, where the goal was to understand how various factors are related to patterns of residual insulin secretion in diabetic children. The data in Figure 10.5 show the age of the children versus the logarithm of their C-peptide concentrations at diagnosis. In the left portion of Figure 10.5 we see the least squares regression line using only the data where age is less than or equal to seven. The straight, nearly horizontal line in Figure 10.5 is the least squares regression line using only the data where age is greater than seven. The corresponding correlations are $r = .635$ and $r = -0.054$, respectively. For the first correlation, which is based on fourteen points, we are able to reject the hypothesis that $\rho = 0$ using the standard Student's T method. So there is empirical evidence suggesting that there is a positive association between age and C-peptide levels, provided attention is restricted to children younger than seven, but for children older than seven it seems that there might be very little or no association at all. (Other modern tools, not covered in this book, lend support to the suggestion that the association changes rather abruptly around the age of seven.) If we use all of the data, meaning we do not divide the data into two groups according to whether a child's age is less than or greater than seven, $r = .4$. Again we reject the hypothesis that $\rho = 0$, in which case the temptation might be to conclude that in general, C-peptide levels increase with age. But as just seen, there are reasons to suspect that this conclusion is too strong and in fact fails to describe the association among children older than seven.

OTHER WAYS PEARSON'S CORRELATION IS USED

There are two additional ways Pearson's correlation is used that should be discussed. The first is that the proportion of variance associated with Y, explained by the least squares regression line and X, is r^2. To elaborate, consider the goal of predicting a child's C-peptide concentration using the data in Figure 10.5, based on the knowledge that the average of all C-peptide concentrations is $\bar{Y} = 1.545$, but without knowing the child's age. Then a reasonable prediction rule is to use $\bar{Y}$. Of course this prediction rule will be wrong in general, and we can measure the overall accuracy of our prediction rule with the sum of the squared differences between $\bar{Y}$ and the C-peptide concentrations we observe. In symbols, we use the sum of the squared errors,

$$(Y_1 - \bar{Y})^2 + \cdots + (Y_n - \bar{Y})^2,$$

to measure how well the sample mean predicts an observation, an idea that was introduced in Chapter 2. For the data at hand, the sum of the squared errors is 1.068. Note that if we divide this sum by $n - 1$, we get the sample variance of the Y values.

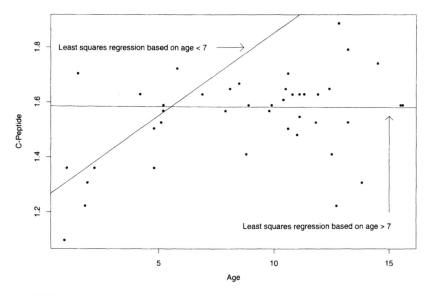

FIGURE 10.5 • These data illustrate a type of curvature that seems to be common: a sudden change in the association. In this particular case there appears to be a positive association for $X < 7$, but no association for $X > 7$.

Another approach to predicting C-peptide concentrations is to use a linear rule and the child's age (X). (We have already seen that using a linear rule might be less than satisfactory for these data, but for illustrative purposes we ignore this concern for the moment.) If we use the least squares regression line, the predicted value of Y, given X, is $\hat{Y} = 0.0153X + 1.41$. So for the first subject in this study whose age is $X_1 = 5.2$, the predicted C-peptide concentration is $\hat{Y}_1 = 0.0153(5.2) + 1.41 = 1.49$, and this individual's actual C-peptide concentration is 1.57. So the residual (the discrepancy between the observed and predicted C-peptide concentration) is $Y_1 - \hat{Y}_1 = 1.57 - 1.49 = .08$. In a similar manner, the age of the second subject is $X_2 = 8.8$ and the residual is $Y_2 - \hat{Y}_2 = -0.13$. Again, we can measure the overall accuracy of our prediction rule with the sum of the squared differences between the observed C-peptide concentrations and their predicted values:

$$(Y_1 - \hat{Y}_1)^2 + \cdots + (Y_n - \hat{Y}_n)^2.$$

For the diabetes data being used here, this sum is equal to 0.898. We see that using our linear rule and the age of the children (X) lowers the sum of the squared errors compared to ignoring X altogether and simply using $\bar{Y}$ to predict C-peptide concentrations. The sum of the squared errors is reduced by $1.068 - 0.898 = 0.17$. The reduction, relative to the sum of the squared errors when a child's age is ignored, is $0.17/1.068 = .16$. This last ratio is just r^2 and is typically called the *coefficient of determination*. That is, r^2 is the proportion of variance accounted for using a least squares regression line and X. But we have already seen that the sample variance is not robust and that its value can be rather unrevealing. So in the current context it is not surprising that r^2 might also be relatively unrevealing.

Finally, it is common to standardize observations in regression, in which case r is the slope of the regression line. Standardizing means that the X scores are changed to

$$Z_x = \frac{X - \bar{X}}{s_x},$$

the Y scores are changed to

$$Z_y = \frac{Y - \bar{Y}}{s_y},$$

and then one fits a least squares regression line to the resulting Z_x and Z_y values. In this case, the regression line is

$$\hat{Z}_y = r Z_x.$$

Note that by transforming the X and Y values in this manner, the sample mean of Z_x, for example, is zero, its sample variance is one, and the same is true for Z_y. A motivation for this approach is that under normality, these Z scores can be interpreted as if observations follow, approximately, a standard normal curve, in which case a probabilistic interpretation of their values might be used. For example, if we assume Z_x has a standard normal probability curve ($\mu = 0$ and $\sigma = 1$), and if an individual has $Z_x = 0$ (meaning that their observed X value is exactly equal to the sample mean, $\bar{X}$), then this subject falls at exactly the middle value of the data, meaning that half of the observed Z_x values are less than zero. Moreover, nonzero Z_x values tell us how many standard deviations away from the mean an observation happens to be. For example, $Z_x = 1$ means that the subject has an X value one standard deviation larger than the mean, and for a normal curve we can attach a probabilistic interpretation to this. In particular, it follows that 84 percent of the subjects have a Z_x value less than one. Having $Z_x = -.5$ means that for the corresponding value for X, 31 percent of the individuals under study have smaller X values. But in Chapter 7 we saw that probabilistic interpretations attached to the standard deviation can be highly misleading even under small departures from normality. Put another way, when dealing with regression, converting to standardized scores does not address problems with nonnormality and can contribute to our misinterpretation of data.

Despite the problems with Pearson's correlation, it has practical utility. For one, it tells us something about the improvement of using a least squares regression line versus using the mean of the Y scores to predict Y, for reasons already noted. In addition, if we reject the hypothesis that Pearson's correlation is zero, using Student's T, we can conclude that the two measures of interest are dependent. If there is in fact independence, then Student's T test provides reasonably good control over the probability of a Type I error. But there is the concern that associations might be missed and even misinterpreted if we insist on limiting our attention to Pearson's correlation. Here we outline and illustrate two general strategies that have been used to develop robust analogs of ρ, but it is stressed that an exhaustive list of every proposed method is not given here. The first of the two strategies first focuses on the X values only and limits the influence of any outlying X values, and then the same is done for the Y values. There are many such methods, but only three are described here. This approach is often quite effective, and it has certain advantages over other strategies that have been used. However, as we shall see, situations can be constructed where this approach can be unsatisfactory. The other general strategy attempts to take into account the overall structure of the data.

THE WINSORIZED CORRELATION

An example of the first strategy, where we guard against both unusual X and Y values, is the Winsorized correlation coefficient. The first step when computing the Winsorized correlation is to Winsorize the observations, but in a manner that leaves paired observations together. An illustration will help clarify what this means.

Imagine that for each of six children we measure hours playing video games per week (X) and reading ability (Y), and we observe

X: 17 42 29 10 18 27
Y: 11 21 47 28 13 25.

The first child plays video games 17 hours per week and has a reading ability of 11. Winsorizing the X values, using 20 percent Winsorizing, means that we first compute $.2n$ and round down to the nearest integer, where n is the number of paired values available to us. Let's label $.2n$, rounded down to the nearest integer, g. Here $n = 6$, so $g = 1$. In the current context, Winsorizing means that we replace the g smallest of the X values with the next smallest value. Because $g = 1$, the smallest value, 10, is increased to the next smallest value, 17. If instead we had $g = 2$, the two smallest values would have been increased to the next smallest. In a similar manner, the g largest values are decreased to the next largest value. Again $g = 1$ in the example, so 42 is pulled down to 29. After Winsorizing only the X values, we now have

X: 17 29 29 17 18 27
Y: 11 21 47 28 13 25.

Notice that we did not change the order of the values so that values not Winsorized remain paired together. For example, the first child played 17 hours of video games per week and has a reading ability of 11, and these two values remained paired together after we Winsorize the X values. But the pair of observations (10, 28) has been changed to (17, 28) after we Winsorize the X values only.

Next we Winsorize the Y values in a similar manner yielding

X: 17 29 29 17 18 29
Y: 13 21 47 28 13 25.

Again, pairs of values not Winsorized remained paired together. For example, for one child we observed the pair (18, 13), and after Winsorizing both the X and Y values, they remain unchanged and are still paired together. Finally, the *sample Winsorized correlation* between X and Y is Pearson's correlation computed with the Winsorized values, which we label r_w to make a clear distinction with r.

Like r, there is a population analog of r_w, which we label ρ_w. It can be shown that when two measures are independent, $\rho_w = 0$. If we can empirically rule out the hypothesis that the population Winsorized correlation is zero, we can conclude that our two measures are dependent. But unlike Pearson's correlation, we can be more certain about what the sign of the Winsorized correlation coefficient is telling us. If the Winsorized correlation is positive, this indicates that there is a positive association among the bulk of the observations. If it is negative, the reverse is true.

One can test
$$H_0 : \rho_w = 0$$

The Winsorized Correlation

using a method very similar to a standard method for testing the hypothesis that Pearson's ρ is zero. The method is based on

$$T = r_w \sqrt{\frac{n-2}{1-r_w^2}}. \tag{10.3}$$

When there is independence, T will have approximately a Student's T distribution with degrees of freedom $h - 2$, where for each variable being studied h is the number of observations not Winsorized. (That is, $h = n - 2g$, where g is $.2n$ rounded down to the nearest integer.)

As an illustration, we again consider the data in Figure 7.10 of Chapter 7, where the goal was to study predictors of reading ability in children. Recall that the least squares regression line was very close to zero and offered no indication that there is an association. This is consistent with Pearson's correlation, which is $r = -0.035$. But as was illustrated in Figure 7.11, there is some reason to suspect that there is indeed an association being masked by outliers when attention is restricted to least squares regression and r. If we replace Pearson's correlation with the Winsorized correlation, it can be seen that $r_w = -.2685$, and straightforward calculations show that Equation 10.3 yields $T = -2.41$. There are seventy-seven pairs of observations. With 20 percent Winsorizing, g is $.2 \times 77$ rounded down to the nearest integer, which is 15. So thirty observations are Winsorized for each of the two variables under study. Consequently, $h = 77 - 30 = 47$, and the degrees of freedom are $47 - 2 = 45$. From tables of Student's T distribution it can be seen that under normality, with 45 degrees of freedom, there is a .95 probability that T will be between -2.01 and 2.01. Because the observed value of T is less than -2.01, we conclude that the two measures are dependent. This illustrates that the choice of method for detecting dependence can be crucial—the Winsorized correlation detects an association, but Pearson's r provides no indication that an association exists.

As already noted, the Winsorized correlation is negative, suggesting that reading ability tends to decrease as the value of our predictor gets large. However, like the standard method based on Pearson's r, our method for testing the hypothesis that the Winsorized correlation is zero is sensitive to both the magnitude of the population Winsorized correlation, ρ_w, and heteroscedasticity. One might argue that although T given by Equation 10.3 indicates that there is dependence, we have not ruled out the possibility that it is heteroscedasticity that caused us to reject $H_0 : \rho_w = 0$ rather than a situation where ρ_w is negative. To add empirical support to the speculation that the Winsorized correlation is indeed negative, we need a method for testing hypotheses about ρ_w

that is sensitive to the value of ρ_w only. It turns out that we can accomplish our goal by switching to the percentile bootstrap method. This simply means we randomly sample n pairs of observations, with replacement, from the n pairs of values obtained from our study and compute the Winsorized correlation coefficient, which we label r_w^*. We repeat this B times, yielding B bootstrap values for the Winsorized correlation. The choice $B = 600$ appears to perform very well in terms of Type I errors. The middle 95 percent of these bootstrap values provides a .95 confidence interval for the population Winsorized correlation coefficient. If this interval does not contain zero, we reject $H_0 : \rho_w = 0$.

Applying this method to the data at hand yields $(-0.491, -0.025)$ as a .95 confidence interval for ρ_w. Because this interval does not contain the value zero, we reject the hypothesis that the population Winsorized correlation is zero. Again we have evidence that our measures are dependent, but unlike before, it is safer to conclude that the population Winsorized correlation is negative. That is, we have stronger evidence that our measure of reading ability is negatively associated with our predictor.

Again, however, care must be taken not to make too strong a statement about the association under study. The bulk of the X values (or predictor values) range between 15 and 89; the six outliers have X values ranging between 133 and 181. So our analysis suggests that over the range of X values between 15 and 89, there is indeed a negative association, but there is some hint that this association might change for X values outside this range. With only six points having X values greater than 100, however, it is difficult to tell.

Why did we Winsorize when computing a correlation? Why not trim instead? The reason has to do with theoretical convenience. When we Winsorize, we can verify that under independence, the population Winsorized correlation is zero. But the same is not necessarily true if we trim instead, so trimming does not lead to a convenient method for establishing dependence. When comparing groups of subjects, however, in terms of some measure of location, trimming is more convenient from a theoretical point of view than when using a Winsorized mean.

SPEARMAN'S RHO

There are two well-known approaches to measuring association that guard against outliers among the X and Y values that are typically covered in an introductory statistics course. They are called *Spearman's rho* and *Kendall's tau*, the first of which is described here. Rather than Winsorize the observations, Spearman's rho converts the observations to so-called *ranks*. To explain, we again consider the data used to illustrate the Winsorized correlation. First consider the X values: 17, 42, 29, 10, 18, 27. The smallest value is said to have rank 1. Here the smallest value is 10, so its rank is 1. The next smallest value has a rank of 2, so here 17 has a rank of 2. We can assign each value a rank by continuing in this fashion. So if we replace the values 17, 42, 29, 10, 18, 29 by their ranks, we get 2, 6, 5, 1, 3, 4. Next we replace the Y values by their ranks, yielding 1, 3, 6, 5, 2, 4. After ranking, our original values

X: 17, 42, 29, 10, 18, 27
Y: 11, 21, 47, 28, 13, 25

become

X: 2, 6, 5, 1, 3, 4
Y: 1, 3, 6, 5, 2, 4.

Spearman's rho, which we label r_s, is just Pearson's correlation computed with the resulting ranks. For the data at hand, $r_s = .2$. Notice that by converting to ranks, we limit the influence of an outlier. For example, if in our illustration the largest X value, 42, were increased to 985, its rank would still be 6, so Spearman's rho remains unchanged. In contrast, Pearson's r drops from 0.225 to -0.11.

It can be shown that like Pearson's correlation, the population value of Spearman's rho (the value we would get if all individuals could be measured) is zero under independence. Moreover, one can test the hypothesis that it is zero using the same method used to test the hypothesis that $\rho = 0$. Illustrations can be found in most introductory textbooks on statistics.

KENDALL'S TAU

Kendall's tau is described with the hypothetical data shown in Table 10.1 where for each individual we have an SAT score and a grade point average (GPA) after four years of college. Notice that for the first two individuals, the SAT scores increase from 500 to 530, and the corresponding GPA scores increase from 2.3 to 2.4. Of course, for a randomly sampled pair of individuals, there is some probability that the student with the higher SAT score will have the higher GPA. Let's label this probability P_G. As is evident, there is some probability that the reverse is true, and we label this P_L. The general idea behind Kendall's tau is to measure the overall tendency for GPA to be higher when the SAT scores are higher, and this is done with

$$\tau = P_G - P_L,$$

the difference between the two probabilities just described. Like Pearson's correlation, it can be shown that τ has a value between 1 and -1 and that it is zero under independence. When $\tau = 1$, for example, it is always true that the student with the higher SAT score has the higher GPA score.

TABLE 10.1 • HYPOTHETICAL DATA ON SAT AND GPA

SAT:	500	530	590	660	610	700	570	640
GPA:	2.3	2.4	2.5	2.6	2.8	2.9	3.3	3.5

Kendall's tau is estimated by determining for each pair of observations whether they are concordant or discordant. In Table 10.1, the first two pairs of observations are said to be concordant because 500 is less than 530 and 2.3 is less than 2.4. That is, the individual with the lower SAT score also has the lower GPA. The pairs (700, 2.9) and (570, 3.3) are said to be discordant because the individual with the higher SAT score has the lower GPA. If we assign the value 1 to a pair of observations that is concordant and the value -1 if the pair is discordant, Kendall's tau is estimated by averaging these values over all possible pairs. For our purposes, the computational details are not important, so no illustration is given. What is important is that Kendall's tau is completely determined by the ordering of the values, so if the largest X value is increased so that it is an outlier, this does not alter the estimate of Kendall's tau. For example, for the data in Table 10.1, the estimate of Kendall's tau is .5. The largest SAT score is 700, and if for the pair of observations (700, 2.9) the value 700 is increased to 1,000,000, this does not alter whether this pair of observations is concordant with any other pair. Consequently, the estimate of Kendall's tau also remains unaltered.

In the illustrations for both Spearman's rho and Kendall's tau, no tied values were used. In Table 10.1, for example, there was only one student who got an SAT score of 500. Had two students gotten a score of 500, then there would be tied values, meaning that an observed value occurred more than once. When using Kendall's tau or any method where observations are converted to ranks, dealing with tied values often requires special care. A book by N. Cliff, nicely summarizes issues and methods for dealing with tied values.

METHODS RELATED TO M-ESTIMATORS

For completeness, it should be mentioned that there are many correlation coefficients related to M-estimators of location. These measures of association are similar to the Winsorized correlation in that they are based on the first general strategy considered here for robustifying Pearson's correlation: Remove or downweight extreme observations among the X values, and do the same for the Y values. But unlike the Winsorized correlation, outliers are identified empirically, an approach that is like the one-step M-estimator of location we saw in Chapter 8. Two such measures of location are the so-called *biweight midcorrelation* and the *percentage bend correlation*. These correlations have practical value when trying to fit a straight line to a scatterplot of points, and the percentage bend correlation offers a good alternative method for establishing dependence, but the computational details are rather involved and are not given here. Chapter 12 indicates where to locate software for applying these correlations, and the bibliographic notes at the end of this Chapter indicate where more information can be found.

A POSSIBLE PROBLEM

So we have outlined three correlation coefficients that guard against unusual X and Y values. Now we illustrate a potential problem with Pearson's correlation that is not corrected by any of these alternative correlations or any other method method based on the general strategy of first eliminating outliers among the X values and then doing the same for the Y values.

Figure 10.6 shows a scatterplot of twenty observations that were generated in the following manner. First, twenty X values were generated on a computer from a (standard) normal probability curve. Then for each X value, a corresponding Y value was generated from a normal curve having mean X and standard deviation one. That is, points were generated that are centered around a straight line having a slope of one. Then two additional points were added, both of which are located at $(2.1, -2.4)$ and appear in the lower-right corner of Figure 10.6.

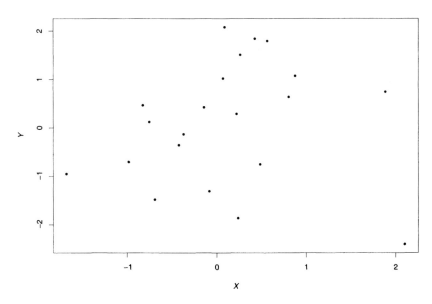

FIGURE 10.6 • An illustration of how points might affect the standard correlation coefficient even when neither the X or Y values are flagged as outliers.

If we focus on the X values only, $X = 2.1$ is not an outlier according to either of the methods described in Chapter 3. In addition, $Y = -2.4$ is not an outlier among all twenty-two Y values. Despite this, Figure 10.6 suggests that the two points at $(2.1, -2.4)$ are unusual, they are somewhat separated from the cloud of the other twenty values, and indeed these two points have a major impact on Pearson's correlation. If we ignore these two points and compute r with the remaining twenty points, we get $r = .443$ and a significance level of .0504. (We are able to reject the hypothesis of a zero correlation if the Type I error probability is set to .0504.) So, there is some indication that X and Y are dependent, which is true by construction. But if we include the two unusual values we get $r = -0.09$ suggesting there is little or no association. There is a positive association for the bulk of the points. One could argue that the two unusual points provide evidence for a weaker association as opposed to a situation where they are not included. Surely this argument has merit, and the decrease in r, when the two outlying points are included, reflects this. However, to conclude there is no association is also misleading. In fact, $X = 2.1$ is the largest of the twenty-two X values, and if we restrict attention to those X values less than 2.1, we again get $r = .44$ giving a strikingly different indication about the association between X and Y.

A POSSIBLE PROBLEM

The more important point is what happens when we switch to one of the alternative measures of correlation just described. If, for example, we use the 20 percent Winsorized correlation, excluding the two points in the lower-right corner of Figure 10.6, we get $r_w = .53$, we reject the hypothesis of a zero correlation, and therefore we conclude that there is dependence. But if we include these two points, the estimated Winsorized correlation drops to $r_w = .24$, and we can no longer reject. (The significance level is .29.) Despite Winsorizing, the two points in the lower-right corner of Figure 10.6 mask an association.

If we switch to Spearman's rho or Kendall's tau, the same phenomenon occurs. For Spearman's rho, ignoring the two unusual points, the correlation is .50, and it is .34 using Kendall's tau. In both cases we reject the hypothesis of a zero correlation (at the .05 level) and correctly conclude that there is an association. But if we include the two unusual points, these correlations drop to .13 and .10, respectively, and we no longer reject.

Why do all of these measures miss the association with the inclusion of the two points in the lower-right corner of Figure 10.6? Roughly, the answer is that these measures of association do not take into account the overall structure of the cloud of points. To elaborate, first notice that the two points $(2.1, -2.4)$ are unusual based on how the other twenty points were generated. To see why, notice that the first twenty Y values were generated from a normal curve having mean X and variance one. When following this method for generating points, if $X = 2.1$, the mean of Y is also 2.1. But recall that for a normal curve, it is highly unusual for a value to be more than two standard deviations from the mean. In the current context, given that $X = 2.1$, it is unlikely that the corresponding Y will be greater than $2.1 + 2(1) = 4.3$ or less than $2.1 - 2(1) = 0.1$. The value $Y = -2.4$ is unusually small, given that $X = 2.1$, because the value -2.4 lies more than four standard deviations away from the mean of Y.

We can graphically illustrate that when $X = 2.1$, having $Y = -2.4$ is unusual relative to the other twenty points using a so-called *relplot*, a bivariate analog of the boxplot, an example of which we have already seen in Figure 7.11. Figure 10.7 shows a relplot for the data in Figure 10.6. (See the bibliographic notes for more information about the relplot.) The inner ellipse contains the central half of the data, and points outside the outer ellipse are declared outliers. So, according to the relplot, the point in the lower-right corner of Figure 10.7 is an outlier, and we have a better visual sense of why this point is unusual.

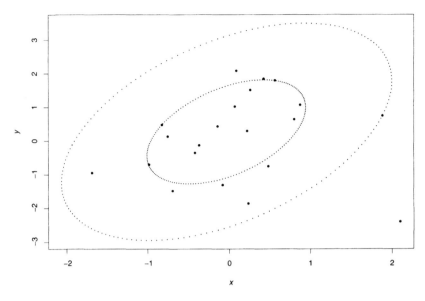

FIGURE 10.7 • A relplot of the data in Figure 10.6. The point in the lower-right corner is outside the outer ellipse, indicating that it is unusually far from the bulk of the points even though its X and Y values are not outliers.

GLOBAL MEASURES OF ASSOCIATION

There are several approaches to measuring the association between two variables that attempt to take the overall structure of a cloud of points into account. Currently, two of these global measures of association are used to detect outliers, but they are not used to test the hypothesis that two or more variables are independent. These are the so-called *minimum volume ellipsoid estimator* and the *minimum covariance determinant estimator* described in this section. In Chapter 11 we will see several regression methods that take into account the overall structure of the points, which can be used to test hypotheses about associations. These regression methods can be used to define new measures of correlation, but the utility of these correlations is still under investigation. For the two measures of association described here, the computational details are quite involved and require a computer, but commercial software is available for applying them. Here the goal is to provide a conceptual basis for these measures. As is the case with so many statistical methods, these measures have great practical value, but some care must be exercised, for reasons to be explained, and not all practical problems have been resolved.

MINIMUM VOLUME ELLIPSOID ESTIMATOR

The relplot in Figure 10.7 provides a convenient graphical tool for explaining one approach to defining a robust measure of correlation. The inner ellipse of Figure 10.7 contains half of the twenty-two points. Notice that this ellipse has some area. Of course, if we draw an ellipse around any other eleven points, its area would generally differ from the area of the inner ellipse in Figure 10.7. One strategy for identifying the central portion of a cloud of points is to search for the ellipse with the smallest area that contains half of the points. The points inside this ellipse can then be used to compute a correlation and a measure of location. In particular, you ignore the points outside the smallest ellipse and merely compute the mean and correlation for the points inside it. This is known as the minimum volume ellipsoid (MVE) estimator. In this manner, points unusually far from the central portion of the cloud of points are eliminated. In fact, the breakdown point of this method is .5.

Finding the smallest ellipse containing half the points is an extremely difficult problem, but effective methods have been devised and are available in some commercial software. (Examples are SAS and S-PLUS.) The computations are lengthy and tedious, so no details are given here. The main point is that this is a viable option now that extremely fast computers are available.

MINIMUM COVARIANCE DETERMINANT ESTIMATOR

Recently, an alternative to the MVE estimator has become practical. The basic idea was proposed by P. Rousseeuw in 1984, and in 1999 he and K. van Driessen published an algorithm for implementing the method. Software for applying the algorithm has been incorporated into version 4.5 of S-PLUS as the function cov.mcd, and into SAS/IML 7 as the function MCD.

To explain the thinking behind this approach to measuring association, we must first describe the notion of a generalized variance. The idea stems from S. Wilks in 1934 and is typically covered in books on the theory of multivariate statistical methods. Again we imagine that for every individual under study, we measure two variables of interest, which we label X and Y. Let σ_x^2 and σ_y^2 be the population variances corresponding to X and Y, respectively, and as usual let ρ be Pearson's correlation. The quantity

$$\sigma_{xy} = \rho \sigma_x \sigma_y$$

is called the *population covariance* between X and Y. The *generalized variance* associated with the variables X and Y is

$$\sigma_g^2 = \sigma_x^2 \sigma_y^2 - \sigma_{xy}^2.$$

(For readers familiar with matrices and multivariate statistics, the generalized variance is the determinant of the covariance matrix.) A little algebra shows that this last equation can be written in a more revealing form:

$$\sigma_g^2 = \sigma_x^2 \sigma_y^2 (1 - \rho^2).$$

The usual estimate of the generalized variance, based on observations we make, is

$$s_g^2 = s_x^2 s_y^2 (1 - r^2).$$

The point is that the generalized variance is intended to measure the overall variability among a cloud of points. Note that s_x^2 measures the variation of X, and of course s_y^2 does the same for Y. The more spread out the X and Y values are, the larger the sample variances. We have also seen that Pearson's correlation, r, is related to how far points are from the regression line around which they are centered. If points are close to the least squares regression line, r^2 tends to be larger than in situations where the points are far from the line. (Smaller residuals result in a larger value for the coefficient of determination, except when points are centered around a horizontal line.) So when the cloud of points is tightly centered around a line, $1 - r^2$ tends to be small, which means that the generalized variance will be small.

Some graphical illustrations might help. The left panel of Figure 10.8 shows one hundred points for which the generalized variance is $s_g^2 = 0.82$. The right panel shows another hundred points, more tightly clustered around the line $Y = X$, and the generalized variance has decreased to $s_g^2 = 0.23$. The left panel of Figure 10.9 shows another hundred points that were generated in the same manner as those shown in the left panel of Figure 10.8, but the variance of the X values was reduced from 1 to 0.5. Now $s_g^2 = 0.20$. In the right panel of Figure 10.9, the points are more tightly clustered together, and the generalized variance has decreased to $s_g^2 = 0.06$.

Now consider any subset of half the points available to us. Each such subset will have a generalized variance. The strategy behind the *minimum covariance determinant* (MCD) estimator is to first select the subset with the smallest generalized variance, the idea being that these points will be clustered together more tightly than any other subset we might consider. Then

GLOBAL MEASURES OF ASSOCIATION

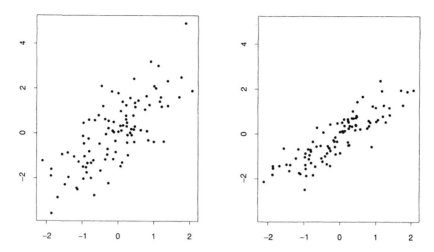

FIGURE 10.8 • Points clustered tightly together tend to have a smaller generalized sample variance, excluding outliers. In the left panel the generalized sample variance is .82; in the right panel it is only .23.

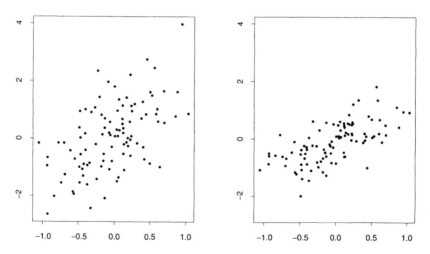

FIGURE 10.9 • Another illustration of how the clustering of points affects the generalized variance. In the left panel it is .20; in the right panel it is .06.

the variances and Pearson's correlation are computed based on this particular subset of points. The resulting value for Pearson's correlation is used as a measure of association for the entire cloud of points. The resulting breakdown point is the same as the MVE estimator, .5, the highest possible value.

An important point is that both the MVE and MCD estimators can be extended to the multivariate case. That is, rather than having two measures for each individual under study, we have p measures, $p \geq 2$, and the goal is to obtain a robust analog of the corresponding variances and covariances that again takes into account the overall structure of the cloud of points. One useful result is an effective outlier detection method that remains valid when points are rotated. (The S-PLUS function outmve in Wilcox, 1997, can be used to check for outliers using the MVE estimator.) Currently, however, the MVE and MCD estimators are not used to test the assumption that two or more variables are independent. There are methods for establishing dependence that take into account the overall structure of the data, some of which are based on regression methods described in Chapter 11. (These regression methods could be used to define new measures of correlation.)

A criticism of the relplot is that it is not a modelfree approach to detecting outliers; data are assumed to be distributed in a certain symmetrical fashion. Note that there is an obvious type of symmetry reflected by the ellipses in a relplot, and violating this assumption of symmetry can affect the conclusions drawn. Even the method used by both the MVE and MCD estimators to detect outliers is not completely modelfree in the sense that properties associated with the normal probability curve play a role in determining whether a point is labeled an outlier. We mention that in 1999, a modelfree approach to detecting outliers for bivariate data, called a *bagplot*, was proposed by P. Rousseeuw, I. Ruts, and J. Tukey. (A closely related approach is the sunburst plot proposed by R. Liu, J. Parelius, and K. Singh in 1999.) Their bagplot visualizes location, spread, correlation, skewness and the tails of data without making assumptions about the data being symmetrically distributed. Moreover, they supply software for implementing the method. A multivariate extension of the bagplot appears possible, but additional technical details must first be resolved.

MORE COMMENTS ON CURVATURE

Although one of the main goals of this book is to explain the strategies used by modern robust methods, another goal is to instill the idea that statistics is not a static area of research, even at the basic level considered here.

For example, the Winsorized correlation, Kendall's tau, and Spearman's rho are able to detect associations that are missed by Pearson's correlation. There is emerging evidence, however, that there is often a curvilinear relationship that is frequently missed by any of these correlations. That is, we test the hypothesis that the correlation is zero and fail to reject because these correlations are not sensitive to the type of curvature being encountered. Efforts are being made to detect dependence in a manner that is more sensitive to a wider range of associations. Published papers strongly suggest that such a method will soon be available and that in applied work, it does indeed have practical value. That is, these new approaches correctly reject in situations where more conventional methods are unlikely to detect an association.

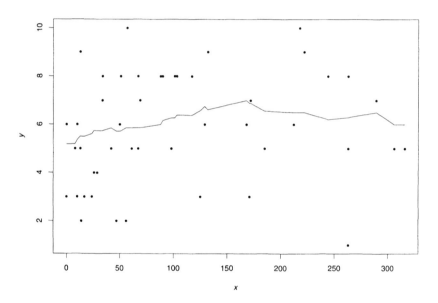

FIGURE 10.10 • Another illustration of why smoothers can be useful in applied work. Both standard and robust methods fail to detect an association, but the smooth suggests that for $X < 150$ there is an association, and this is confirmed by both robust and traditional techniques.

Also, the utility of smoothers when dealing with curvature cannot be stressed too strongly. Figure 10.10 shows a scatterplot of data from a study on the effects of aggression in the home. (The data were generously supplied by A. Medina.) Pearson's correlation, and the robust methods described here fail to detect any association. However, the smooth in Figure 10.10 suggests that there is a positive association up to about $X = 150$, after which there seems to be little or no association at all. If we restrict the range of X and consider only values less than 150, Pearson's correlation and various robust methods indicate that an association exists. (These methods are significant at the .05 level.)

A SUMMARY OF KEY POINTS

- Pearson's correlation is a useful tool for summarizing and detecting associations, but it can be relatively unrevealing because at least five features of data (summarized at the end of Section 10.1) influence its magnitude. (In psychometrics, reliability also influences the magnitude of r.)

- Smoothers provide a useful addition to the many tools one might use to study and detect curvature. In some cases smoothers indicate that there is an association between X and Y over some range of X values, but outside this range, the association disappears.

- One way of reducing the effects of outliers on r is to downweight extreme values among the X values and do the same for the Y values. The Winsorized correlation, Spearman's rho, and Kendall's tau are three examples of this approach. These correlations can improve our ability to detect associations and describe how the bulk of the observations are associated, but problems persist. Global measures of association have been proposed and have practical value when trying to detect outliers among multivariate data.

- Student's T can be used to test the hypothesis of zero correlation using any of several correlation coefficients. All such tests assume homoscedasticity. Tests that allow heteroscedasticity can be performed using an appropriate bootstrap technique.

BIBLIOGRAPHIC NOTES

The MCD estimator was first proposed by Rousseeuw (1984), and the recent algorithm for implementing it is described in Rousseeuw and van Driessen (1999). For a more comprehensive discussion of the MVE estimator and related methods, see Rousseeuw and Leroy (1987). Rousseeuw and van Zomeren (1990) describe how the MVE estimator can be used to detect outliers in multivariate data, and a similar strategy can be used with the MCD estimator. For software and a description of the bagplot, see Rousseeuw, Ruts, and Tukey (1999). Liu, Parelius, and Singh (1999) describe a sunburst plot that is similar to the bagplot. For more details about the Winsorized correlation, see Wilcox (1997). For a useful alternative to the correlations described here, see the description of the percentage bend correlation in Wilcox (1997). For information about how to handle tied values when using Kendall's tau, see Cliff (1996). For a method that deals with curvature when trying to detect an association, see Wilcox (in press a).

CHAPTER *11*

ROBUST REGRESSION

As previously indicated, least squares regression suffers from several practical problems. A single unusual point can mask an important and interesting association, and least squares regression can give a distorted view of how the bulk of the observations are related. Even when the outcome measure (Y) has a normal distribution, heteroscedasticity can result in relatively low power when testing the hypothesis that the slope is zero. That is, heteroscedasticity can mask an association because the variance (or squared standard error) of the least squares estimator can be very high compared to other estimators one might use. Nonnormality makes matters worse, and the conventional method for computing a confidence interval for the slope can be highly inaccurate.

It would be convenient if a single regression method could be identified and recommended over all other methods one might use. Unfortunately, the perfect estimator has not been found and appears not to exist. What is best in one situation might not be best in another. A complicating factor is that several criteria are used to judge any method, and even if one method dominates based on one criterion, it might not dominate when using another. In terms of achieving the smallest standard error, estimator A might beat method B in some situations but not others.

We can, however, classify regression methods into one of two groups: those that perform relatively well over a reasonably wide range of situations and those that do not. The least squares estimator falls in the latter category. It performs well when there is both normality and homoscedasticity, but otherwise it can be extremely unsatisfactory. If our only concern is the probability of a Type I error when testing the hypothesis of independence, least squares performs reasonably well. But if we want to detect an association and provide an accurate description of what this association is, least squares can fail miserably.

There are several methods that compete well with least squares when there is normality and homoscedasticity. Simultaneously, they can be strikingly more accurate when there is nonnormality or heteroscedasticity. The best advice at the moment is to be familiar with several of these methods and to know something about their relative merits. The goal in this chapter is to describe a few of them and mention where information about other methods can be found. The computational details for some of the methods are easily described, but for others they are rather lengthy and complicated. The focus here is on understanding the strategy behind the various methods proposed and on describing why one method might be preferred over another. Software is available for applying all the methods described, and for the computational details associated with the more complicated methods, interested readers can refer to the bibliographic notes at the end of this chapter. Initially the focus is on the simple case, where there is one predictor. In Section 11.12 the relative merits of the methods are summarized and comments about handling multiple predictors are made.

Some methods are included in this chapter, not because they have good properties, but for the sake of completeness. Some approaches might seem reasonable, but there may be practical problems that are not immediately obvious.

As a final introductory remark, although no single method is always perfect, a few methods can be identified that might be given serious consideration for general use. Generally, it is suggested that multiple tools be used when studying associations. Regression is a very difficult problem that needs to be considered from several vantage points.

THEIL-SEN ESTIMATOR

As noted in Chapter 2, for a scatterplot of n points, any two distinct points can be used to determine a slope. For the data in Table 2.1, which listed five points, there are ten pairs of points one could use to compute a slope. These ten slopes, written in ascending order, are

$$-349.19, 133.33, 490.53, 560.57, 713.09, 800.14, 852.79,$$
$$957.48, 1185.13, 1326.22. \quad (11.1)$$

In Section 2.7.1 it was pointed out that least squares regression corresponds to taking a weighted average of all these slopes. But instead of taking a weighted average of the slopes, what if we take the median of all the slopes? For the ten pairs of points considered here, the median of the corresponding slopes is 757.6 and represents a reasonable estimate of the population slope, the slope we would get if infinitely many observations could be made. Letting b_{ts1} represent the median of the slopes we get for any n points, the intercept can be estimated with

$$b_{ts0} = M_y - b_{ts1} M_x,$$

where M_y and M_x are the medians associated with X and Y, respectively. This strategy for estimating the slope and intercept was formally studied by H. Theil and was later extended by P. K. Sen and is now called the *Theil-Sen estimator*. (In actuality, Theil developed an estimate of the slope based on Kendall's tau, which turned out to be tantamount to using the median of the slopes associated with all pairings of points.) Two hundred years ago, the Theil-Sen estimator would have been a natural approach to consider, and it is in fact similar to Boscovich's method covered in Chapter 2. But at the time it could not have competed very well with least squares for at least two practical reasons. First, as the number of points available to us increases, the number of slopes that must be computed soon exceeds what can be done in a reasonable manner without a computer. Second, a convenient method for measuring the precision of the estimated slope (computing a confidence interval) was not available. Because a relatively simple method for computing a confidence interval for the slope had been derived when using least squares, it had to be very appealing at the time. Simultaneously, it was difficult—and perhaps impossible—to appreciate the practical problems with least squares that would be revealed during the latter half of the twentieth century. Today, thanks in part to fast computers, the practical problems associated with the Theil-Sen estimator can be addressed. Computing the median of the slopes of all pairs of points is easily done on a desktop computer, and an accurate confidence interval can be computed even when there is nonnormality and heteroscedasticity.

There are at least three practical advantages associated with the Theil-Sen estimator. The first is that the breakdown point is .29, which is high enough (in most cases) to avoid misleading results due to outliers. Second, the standard error of the Theil-Sen estimator competes fairly well with the least squares estimator when there is both normality and homoscedasticity. That is, in the situation where least squares performs best according to theory, it offers only a slight advantage compared to Theil-Sen. Third, even under normality, if there is heteroscedasticity, the Theil-Sen estimator can have a much smaller standard error, meaning that on average, it is a more accurate estimate of the true slope. In some cases the standard error of the Theil-Sen estimator can be hundreds of times smaller than the least squares estimator!

In terms of achieving a small standard error, the success of the Theil-Sen estimator might seem rather surprising based on comments about the median given in previous chapters. Why not use less trimming? For example, rather than taking the median of all the slopes, why not compute a 20 percent trimmed mean instead? The reason is that if we were to create a boxplot of the slopes corresponding to all pairs of points, we would see that outliers are common. It is not surprising to find outliers because when we compute a slope based on only two points, it is easy to imagine situations where two specific points give a wildly inaccurate estimate of the true slope. What is perhaps less obvious is just how often this happens. The main point is that consideration has been given to replacing the median with the 20 percent trimmed mean for the problem at hand, and all indications are that the median is preferable for general use.

To compute a confidence interval for both the slope and the intercept, when employing the Theil-Sen estimator, one merely applies a version of the percentile bootstrap that allows heteroscedasticity. More precisely, first generate a bootstrap sample by resampling, with replacement, n pairs of points from the n pairs available to us. This is how we began when using the bootstrap with the least squares regression line, as described in Chapter 6. Repeat this process B times and take the middle 95 percent of the resulting bootstrap estimates of the slope as a .95 confidence interval for the true slope. That is, use the same bootstrap method employed with the least squares estimator, making no special adjustments when the sample size is small. (With least squares, an adjustment is made for $n < 250$.) A confidence interval for the intercept can be computed in a similar manner. When using the Theil-Sen estimator, it has been found that $B = 599$ gives good results for a wide range of situations when computing a .95 confidence interval.

As an illustration, consider again Boscovich's data listed in Table 2.1. A .95 confidence interval for the slope, based on the Theil-Sen estimator, is (70.7, 1185.1). Note that this confidence interval does not contain zero, so we would reject the hypothesis of a zero slope. If we use least squares instead, in conjunction with the bootstrap method, the .95 confidence interval is (−349.2, 1237.9), which is distinctly different from and quite a bit longer than the confidence interval using Theil-Sen. Indeed, the least squares method cannot even reject the hypothesis of a zero slope.

As another example, consider the star data shown in Figure 6.6 of Chapter 6. One might argue that merely looking at a scatterplot suggests that points in the left portion of the scatterplot are unusual and might affect our conclusions about how the two variables are related. Suppose we eliminate the outliers in the left portion by considering only X values greater than 4.1. (If we eliminate outlying Y values and apply standard methods, we get the wrong standard error for reasons similar to those described in Chapter 9.) Now a scatterplot of the points appears as shown in Figure 11.1. The top line is the Theil-Sen regression line; the bottom line is the least squares regression line. A bootstrap .95 confidence interval for the slope, using the least squares estimator, is (1.63, 4.1). Using the Theil-Sen estimator instead, the .95 confidence interval is (2.1, 4.2). Again we get a shorter confidence interval. Moreover, the Theil-Sen estimator suggests that a slope of 2 or less should be ruled out, but with the least squares estimator we cannot reject the hypothesis that the slope is as small as 1.7, the only point being that it can make a practical difference which method is used.

It should be noted that if we use the conventional confidence interval for the slope, based on the least squares estimator and Student's T, situations arise in applied work where we get a slightly shorter confidence interval than with the bootstrap with the Theil-Sen estimator. This is not a very compelling reason for using the former method, however, because its probability coverage can be very poor. If we compute a confidence interval that actually has only a .6 probability of containing the true slope, it will tend to produce a shorter confidence interval than a method that has a .95 probability coverage. Of course, if our intention is to have .95 probability coverage, producing a confidence interval with .6 probability coverage is unsatisfactory even if its length is shorter.

Despite possible problems with probability coverage when using Student's T, it gives (1.9, 4.1) as a .95 confidence interval for the slope based

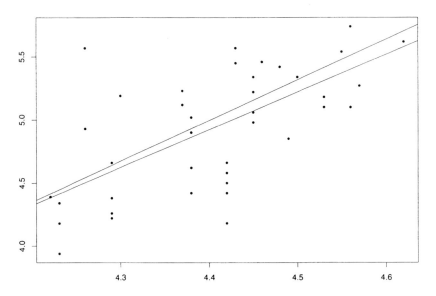

FIGURE 11.1 • A scatterplot of star data with X values less than 4.1 eliminated. The top and bottom lines are the Theil-Sen regression line and the least squares line, respectively. The two lines appear to be reasonably similar, but the confidence interval for the Theil-Sen line is substantially shorter.

on the data shown in Figure 11.1. So although there is some possibility that the actual probability coverage is less than .95, the Theil-Sen estimator gives a shorter confidence interval, albeit by a small amount.

REGRESSION VIA ROBUST CORRELATION AND VARIANCES

The first section of Chapter 10 noted that the least squares estimate of the slope can be written as

$$b_1 = r \frac{s_y}{s_x}.$$

This suggests a simple method for getting a more robust estimator: Replace Pearson's correlation r and the standard deviations with measures of correlation and variance that are robust to outliers. For example, we might replace Pearson's correlation with r_w, the Winsorized correlation, and replace s_y and s_x with the Winsorized standard deviations, s_{wy} and s_{wx}. There are, in fact, many variations of this approach that have been considered, some of which might be better in general than Winsorizing. For example, one might use a measure of correlation and variation related to M-estimators of location. One particular choice that seems interesting is the so-called biweight measure of association and dispersion. A possible advantage of this approach is a higher breakdown point, but currently there is no formal proof as to whether 20 percent Winsorizing would have a lower breakdown point. To complicate matters, there are situations where Winsorizing has a much smaller standard error than using the biweight, and there are situations where the reverse is true. Consequently, the lengthy details of how to use a biweight measure of association are not covered here, but it is certainly not being suggested that this approach be ruled out. Yet another possibility is to use a rank-based measure of association such as Kendall's tau.

Another complication is that in some—but not all—cases, simply replacing r, s_x, and s_y with robust analogs can result in an estimation problem that has considerable practical concern. To explain, first consider the least squares regression estimate of the slope, b_1. If we were to repeat a study billions of times (in theory, infinitely many times), each time using n randomly sampled points to compute the least squares estimate of the slope, the average of these billions of estimates would be β_1, the true (population) slope. In this case, b_1 is said to be an *unbiased* estimator. More formally, the expected value of the least squares slope is the population slope, the slope we would get if all individuals of interest could be measured. In symbols, $E(b_1) = \beta_1$. If we use a Winsorized correlation and standard deviation instead, the resulting estimate of the slope is approximately unbiased provided the errors (or residuals) are homoscedastic. But if they are heteroscedastic, an extreme amount of bias is possible. Fortunately, with the aid of a computer, there is an effective (iterative) method for correcting this problem. The bibliographic notes indicate where to find more details.

L_1 REGRESSION

Recall from Chapter 2 that when working with measures of location, squared error leads to the sample mean and absolute error leads to the median. The median has a high breakdown point, so to avoid the effects of outliers in regression, a natural strategy is to use absolute errors. In particular, imagine we observe n pairs of observations, which we label $(X_1, Y_1), \ldots, (X_n, Y_n)$.

Then for any regression line $\hat{Y} = b_1 X + b_0$ the corresponding residuals are

$$r_1 = Y_1 - \hat{Y}_1 = Y_1 - b_1 X_1 - b_0$$
$$r_2 = Y_2 - \hat{Y}_2 = Y_2 - b_1 X_2 - b_0$$
$$\vdots$$
$$r_n = Y_n - \hat{Y}_n = Y_n - b_1 X_n - b_0.$$

So for a specific choice for the slope and intercept, r_1, for example, represents the discrepancy between Y_1, what we observe, and $\hat{Y}_1$, its predicted value. The least squares approach to choosing the slope and intercept measures the overall accuracy of the prediction rule $\hat{Y}$ with $r_1^2 + \cdots + r_n^2$, the sum of the squared residuals. Suppose instead that we measure the overall accuracy of the prediction rule with $|r_1| + \cdots + |r_n|$, the sum of the absolute residuals. If we choose the slope (b_1) and the intercept (b_0) to minimize the sum of the absolute residuals, we get what is called the L_1 or the *least absolute value* estimate of the slope and intercept. This approach predates the least squares approach by at least fifty years.

The good news about the least absolute value estimator is that it protects against outliers among the Y values. But a concern is that it does not protect against outliers among the X values. In fact, its breakdown point is only zero. Figure 11.2 illustrates the problem. The Theil-Sen estimator captures the association among the bulk of the observations, but the L_1 estimator does not. Another criticism is that L_1 gives too much weight to observations with small residuals. Examples can be constructed where there are outlying X values, yet L_1 gives a good fit to most of the points. However, because it can be highly misleading, it seems that it should be given relatively little consideration compared to some of the alternative methods described in this chapter. For these reasons, it is included here for completeness, but additional details are omitted. Despite its problems, some of the more successful estimators use the L_1 method as a step toward a more effective approach, including methods that have a high breakdown point.

LEAST TRIMMED SQUARES

Another approach is to simply ignore or trim the largest residuals when judging how well a particular choice for the slope and intercept performs. For example, imagine we have n points, and as usual we let $r_1, \ldots, r_n$ represent

LEAST TRIMMED SQUARES

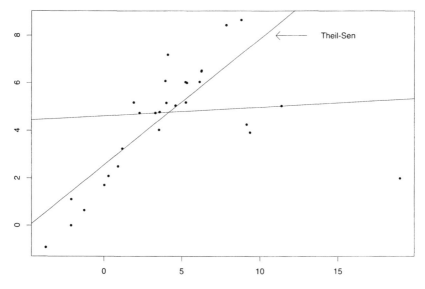

FIGURE 11.2 • The least absolute value regression line misses the positive association for the bulk of the points, but the Theil-Sen line captures it.

the residuals corresponding to some choice for the slope and intercept, which we again label b_1 and b_0. Next, order the residuals and label them $r_{(1)} \leq \ldots \leq r_{(n)}$ so that $r_{(1)}$ is the smallest residual, $r_{(2)}$ is the next smallest, and $r_{(n)}$ is the largest. Finally, let $h = .75n$. Rather than judge the performance of our prediction rule $\hat{Y} = b_1 X + b_0$ based on the sum of the squared residuals, suppose we judge it based on the sum of the h smallest squared residuals instead. So, for example, if $n = 100$, then $h = 75$ and we would use the sum of the seventy-five smallest squared residuals, $r_{(1)}^2 + r_{(2)}^2 + \cdots + r_{(75)}^2$, to judge how well $\hat{Y}$ provides a good fit to data. More generally,

$$r_{(1)}^2 + r_{(2)}^2 + \cdots + r_{(h)}^2 \tag{11.2}$$

is used to judge fit. The strategy behind the *least trimmed squares* estimator is to choose b_1 and b_0 to minimize the sum given by Equation 11.2. In effect, we remove the influence of the $n - h$ largest residuals by trimming them.

If we set $h = .75n$, the breakdown point of the least trimmed squares estimator is $1 - .75 = .25$. So about 25 percent of the points can be outliers without completely destroying the estimator. Of course, we need not set $h = .75n$. We could use $h = .8n$ instead, in which case the breakdown point is

$1 - .8 = .2$. With $h = .5n$ the breakdown point is .5, but for technical reasons the smallest value for h that should be used is $n/2$ rounded down to the nearest integer, plus one. For example, if $n = 101$, $101/2 = 50.5$, rounding down yields 50, and adding one gives 51. So the smallest value for h that would be used is 51, which is close to a breakdown point of .5. (For the general case where there are p predictors, the smallest h used is $[n/2] + [(p + 1)/2]$, where $[n/2]$ and $[(p + 1)/2]$ are $n/2$ and $(p + 1)/2$ rounded down to the nearest integer.)

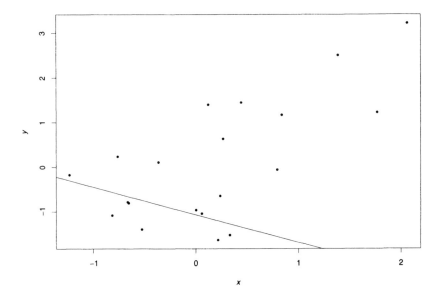

FIGURE 11.3 • A low breakdown point can be disastrous, but there are situations where the highest breakdown point might also be unsatisfactory. In this case the highest breakdown point for the least trimmed squares estimator misses the true association completely, but with a breakdown point of .2 or .25 it is detected.

To add perspective, it should be noted that when sample sizes are small, simply using a regression estimator with a high breakdown point is no guarantee that disaster will be averted. We need to consider more than just the breakdown point when choosing an estimator. As an illustration, Figure 11.3 shows a scatterplot of twenty points that were generated in the same manner as the points in Figure 10.6 of Chapter 10. In particular the points were generated from a regression line having a slope of 1 and an intercept of 0, and both the X and Y values have a normal distribution. If we use the maximum possible breakdown point when applying the least trimmed squares estimate of the slope (using the S-PLUS built-in function ltsreg), the estimate of the slope is -0.62. In this particular case, by chance, we get a highly inaccurate estimate of the true slope. If, however, we use $h = .8n$, so that the breakdown point is .2, the estimate of the slope is 1.37, and for $h = .75n$ it is 0.83. So a lower breakdown point yields a much more accurate estimate in this particular case. As for Theil-Sen, it yields an estimate of 1.01, which happens to be even more accurate than the least squares estimate, 1.06. Of course, this is not a compelling reason to prefer Theil-Sen over the least trimmed squares estimator. The only point is that a highly erroneous estimate is possible using the highest possible breakdown value of the least trimmed squares estimator, at least with a sample size of only twenty.

It is not being suggested that all estimators with a high breakdown point give a highly erroneous estimate of the slope based on the data in Figure 11.3. The least median of squares estimator also has a breakdown point of .5, and for the data in Figure 11.3, the estimated slope is 1.36. But as we shall see, the least median of squares estimator does not dominate the least trimmed squares estimator.

To elaborate a bit on how much trimming to do when using the least trimmed squares estimator, the process used to generate the data in Figure 11.3 was repeated one hundred times. The left portion of Figure 11.4 shows a boxplot of the one hundred estimates based on the highest possible breakdown point, and the right boxplot shows the results using a breakdown point of .2. Notice that with a breakdown point of .2, the slope estimates are more tightly centered around the correct value of one. In this particular case, we are better off, in general, using a lower breakdown point. But examples can be constructed where the opposite is true. What seems to suffice in most situations is a breakdown point around .2 or .25, but the only certainty is that exceptions might be encountered. This suggests that in the preliminary stages of analysis, perhaps multiple methods, having a reasonably high breakdown point, should be considered. If large discrepancies are encountered, trying to understand why using graphical or other diagnostic tools is helpful.

Like all of the regression methods covered in this chapter, a percentile bootstrap method can be used to compute confidence intervals. If we apply this method to the data in Figure 11.3, using the highest possible breakdown point, the .95 confidence interval is $(-1.4, 2.1)$.

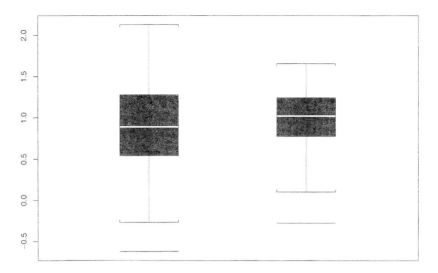

FIGURE 11.4 • Boxplots illustrating the accuracy of the least trimmed squares estimator with different breakdown points. The true slope being estimated is one. On the left is a boxplot of one hundred estimates using the highest breakdown point, and on the right a breakdown point of .2 was used. Notice that with a breakdown point of .2, the estimates are more tightly centered around the correct slope.

LEAST TRIMMED ABSOLUTE VALUE

There is a slight variation of the least trimmed squares estimator that should be mentioned. Rather than use the squared residuals, use the absolute values of the residuals. Letting h be defined as in Section 11.4, we replace Equation 11.2 with

$$|r_{(1)}| + |r_{(2)}| + \cdots + |r_{(h)}|, \tag{11.3}$$

and the goal is to find the values for b_1 and b_0 that minimize Equation 11.2. Given h, the breakdown point is the same as the least trimmed squares estimator. There is weak evidence that with a fixed breakdown point, the least trimmed squares estimator tends to be more accurate on average than using the least trimmed absolute residuals used here. But with certain types of heteroscedasticity, the reverse is true. This issue is in need of a more detailed analysis.

LEAST MEDIAN OF SQUARES

Yet another approach to regression is to choose the slope and intercept that minimizes the median of the squared residuals. That is, any choice for the slope and intercept is judged based on the median of $r_1^2, \ldots, r_n^2$, and the goal is to choose the slope and intercept that minimize this median. Again the breakdown point is .5, the highest possible value. As an estimator of the slope, the least median of squares approach is not considered to be very competitive compared to most other estimators. The least trimmed squares estimator, for example, using a breakdown point of .5, has been found to have better mathematical properties and is typically a more accurate estimate of the true slope. But as previously illustrated, when the sample sizes are small, using the highest breakdown point can result in a highly misleading result. Moreover, it is possible, though perhaps rare, to encounter data where the least median of squares estimator is considerably more accurate. As indicated in Figure 11.3, the least median of squares estimate of the slope (using the S-PLUS function lmsreg) gives a much more accurate estimate of the correct slope than using least trimmed squares with the highest breakdown point. This is not an argument for preferring the least median of squares method over the least trimmed squares approach. This merely illustrates that caution must be exercised when using either method.

REGRESSION OUTLIERS AND LEVERAGE POINTS

Before continuing with our description of regression estimators, we need to digress and describe what are called regression outliers. *Regression outliers* are points with unusually large residuals based on a regression line that has a reasonably high breakdown point. That is, if the bulk of the points under study follow a linear pattern, regression outliers are points that are relatively far from the line around which most of the points are centered. Figure 11.5 shows a scatterplot of the logarithm of brain weight and body weight of twenty-eight animals. The three circled points in the right portion of Figure 11.5 appear to be relatively far from the line around which most of the points are centered. (The regression line in Figure 11.5 is the least median of squares regression line.)

One proposed rule for deciding whether a point is a regression outlier is as follows. Determine the least median of squares regression line and then

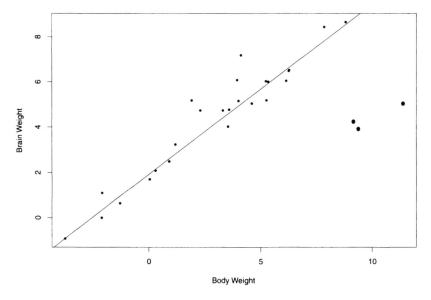

FIGURE 11.5 • The three circled points on the right side of this scatterplot are regression outliers: points relatively far from the line around which the bulk of the points are centered.

compute the residuals. Label the median of the residuals M_r. Then the point (X_i, Y_i), having a residual r_i, is labeled a regression outlier if $|r_i|/\hat{\tau} > 2.5$, where

$$\hat{\tau} = 1.4826 \left(1 + \frac{5}{n-p-1}\right) \sqrt{M_r}$$

and p is the number of predictors. Using this method, the three circled points in the right portion of Figure 11.5 are declared regression outliers. In this particular case twenty-five of the animals are mammals, none of which is declared an outlier. The three outliers correspond to three different types of dinosaurs.

In regression, unusually large or small X (predictor) values are called *leverage points*. Of course, we can check for leverage points using one of the robust methods described in Chapter 3. But in regression, it helps to make a distinction between two types of leverage points: good and bad. A *good leverage point* is a point that is unusually large or small among the X values but is not a regression outlier. That is, the point is relatively removed from the bulk of the observations but reasonably close to the line around which most of the points are centered. A good leverage point is shown in the upper-right portion of Figure 11.6. A *bad leverage point* is a leverage point that has an unusually large residual corresponding to some robust regression line having a reasonably high breakdown point. The idea is that a bad leverage point is a point situated far from the regression line around which the bulk of the points are centered. Said another way, a bad leverage point is a regression outlier that has an X value that is an outlier among all X values as well. The point labeled a bad leverage point in Figure 11.6 has an X value that is an outlier among all X values under study, and it is relatively far removed from the regression line in Figure 11.6, which is the least median of squares line.

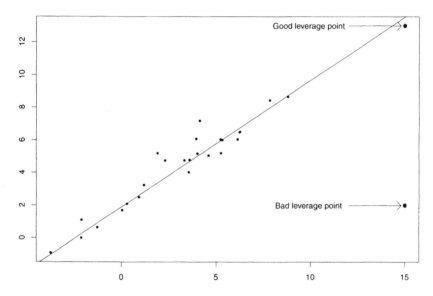

FIGURE 11.6 • Examples of both good and bad leverage points.

What makes a good leverage point good? And what makes a bad leverage point bad? Bad leverage points can grossly affect your estimate of the slope if you use an estimator with a small breakdown point. For example, the slope of the regression line in Figure 11.6 is 0.77 and clearly gives a good indication of how the bulk of the observations are associated. But the slope of the least squares estimate is 0.5 and gives a poor fit relative to the majority of the points because a single leverage point can have an inordinate influence on the least squares estimate. A good leverage point has a limited effect on giving a distorted view of how the majority of points are associated. And when using least squares, it can have the advantage of decreasing the standard error. To see why, notice that from Chapter 4, the squared standard error of the least squares estimated slope is inversely related to the sample variance of the X values, s_x^2. That is, the more spread out the X values happen to be, the more this lowers the estimated standard error. Because outliers among the X values can increase s_x^2 substantially, the variance of the least squares estimate of the slope is decreased when there are outliers among the X values. Even bad leverage points lower the standard error, but this can be more than offset by the inaccurate estimate of the slope that can result.

M-ESTIMATORS

As indicated in Chapter 8, M-estimators of location include a large class of estimators that are distinguished by how we measure error. These same measures of error can be used in the more general situation considered here. But because there are so many versions of M-estimators and so many technical issues, a comprehensive description of this approach to regression is impossible without getting bogged down in technical issues. Moreover, the computational details associated with M-estimators are rather involved and tedious. However, in terms of software, some of the better M-estimators are the easiest to use, particularly when there are multiple predictors. The goal in this section is to outline the method and encourage interested readers to refer to more advanced books on this interesting topic.

Rather than choose the slope and intercept to minimize the sum of the squared residuals, the strategy used by least squares, M-estimators minimize the sum of some convenient function of the residuals instead. In symbols, we choose the slope (b_1) and intercept (b_0) to minimize

$$\xi(r_1) + \xi(r_2) + \cdots + \xi(r_n), \tag{11.4}$$

where ξ is some function chosen to guard against outliers and heteroscedasticity. The choice $\xi(r) = r^2$ leads to least squares and $\xi(r) = |r|$ leads to the L_1 estimator already described. Again, Huber's measure of error has practical advantages, but there are new problems that must be taken into account.

Chapter 8 indicated that when dealing with M-estimators of location, often a measure of scale must be incorporated into how the measure of location is defined (in order to achieve scale equivariance). For the more general situation considered here, again a measure of scale must be used when adopting one of the many interesting methods for measuring error. Specific choices for a measure of scale have been studied and recommendations can be made, but the details are not covered here. What is more important is getting a general sense about the strategy behind M-estimators and understanding their relative merits. (As usual, computational details can be obtained from books listed in the bibliographic notes.)

A practical concern is that some versions of M-estimators perform rather poorly in terms of having a small standard error when there is heteroscedasticity. But there is a method for dealing with this issue using what are called *Schweppe weights*. To complicate matters, there are several versions of Schweppe weights, and not all versions can be recommended. Here it is merely remarked that at least one version, yielding what is called an *adjusted M-estimator*, appears to have great practical value; it has the highest possible breakdown point, .5, and its standard error competes well with both least squares and the other estimators described in this chapter. Part of the strategy behind the so-called adjusted M-estimator is to first check for bad leverage points. These points are ignored when estimating the slope and intercept. In addition, when judging how well a choice for the slope and intercept performs based on Equation 11.3, adjustments are made that result in a relatively small standard error when there is heteroscedasticity.

Again, this is not to suggest that this estimator should be routinely used to the exclusion of all others. It is merely being suggested that it be included in the collection of techniques one might consider. Although complete details are not given here, easy-to-use software is available, and this approach to regression seems to be one of the more effective methods currently available. The relative merits of all the estimators described here will be discussed in Section 11.10.

As usual, computing confidence intervals and testing hypotheses can be done with the bootstrap method already described in conjunction with the Theil-Sen estimator. All indications are that relatively accurate results can be obtained with fairly small sample sizes, even under extreme departures from normality and when there is an extreme amount of heteroscedasticity.

There is an important point that cannot be stressed too strongly. Even among the robust regression estimators listed in this chapter, the choice of which method to use can make a practical difference. In fact, even if two robust estimators give nearly identical regression lines, there might be a practical reason for preferring one estimator over another when computing a confidence interval or testing hypotheses. As an illustration, Figure 11.7 shows the average 1992 salaries of assistant professors at fifty universities versus the average salaries of full professors. Included are the regression lines based on Theil-Sen and the adjusted M-estimator, and as is evident, the two lines are similar. If we generate B bootstrap estimates of the slope using, say, the Theil-Sen estimator, the variance of these bootstrap values estimates the squared standard error of the Theil-Sen estimator, and the square root estimates the standard error. If we replace the Theil-Sen estimator with the adjusted M-estimator, we get an estimate of the standard error of the adjusted M-estimator instead.

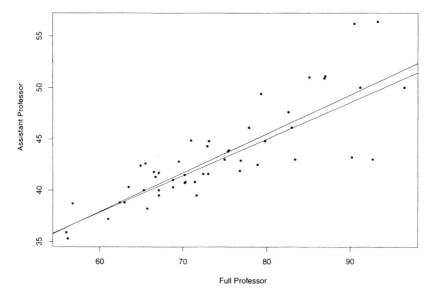

FIGURE 11.7 • The two regression lines are the Theil-Sen and the adjusted M-estimator. Both robust estimates are similar, but in this particular case Theil-Sen produces a substantially shorter confidence interval.

For the data in Figure 11.7, the estimated standard error of the Theil-Sen estimator is 0.047. As for the adjusted M-estimator, the estimate is 0.094, nearly twice as large, suggesting that the Theil-Sen estimator is more accurate on average. This difference in the standard errors is reflected by the corresponding .95 confidence intervals. Using Theil-Sen the .95 confidence interval is (0.27, 0.47) versus (0.11, 0.50) using the adjusted M-estimator. In this case it appears that Theil-Sen provides a more accurate estimate of the slope in the sense that its confidence interval yields a narrower range of possible values. This illustrates that using Theil-Sen rather than the adjusted

M-estimator can have practical value in some situations because shorter confidence intervals can mean more power. But this one example does not provide even a remotely convincing argument that Theil-Sen should be routinely used to the exclusion of the adjusted M-estimator—it merely indicates that the choice of an estimator can make a difference in the conclusions reached.

THE DEEPEST REGRESSION LINE

Recently there has been interest in yet another approach to regression that looks very promising. It is based on a mathematical method for quantifying how deeply a regression line is embedded in a scatterplot of points. Any line can be given a numerical value, called its *regression depth*, that measures its embeddedness. The regression strategy is to consider all pairs of distinct points, determine the slope and intercept for each such pair, and then compute the regression depth corresponding to this line. Having done this for all pairs of points, one then selects the line with the highest regression depth. If several lines have the highest depth, their slopes and intercepts are averaged. This approach has a breakdown point of .33 when X and Y have a linear association. Recommending this method is a bit premature, so no computational details are given, but early results suggest that it is something applied researchers might want to consider in the near future.

RELATIVE MERITS AND EXTENSIONS TO MULTIPLE PREDICTORS

Finding a method that generally beats least squares has been accomplished; many such methods are available today. But choosing which method to use routinely is another matter. When there is a single predictor, the Theil-Sen estimator satisfies the basic requirements of having a reasonably high breakdown point and comparing well to least squares regression in terms of achieving a relatively low standard error. At a minimum, any alternative to least squares should provide a striking advantage in some cases, and Theil-Sen meets this requirement. Another convenience is that Theil-Sen is easy to explain. But two points should be stressed. First, situations can be found where alternative methods offer a substantially lower standard error, but there are also situations where the reverse is true. The only way to know whether switching to an alternate estimator makes a difference is to simply try the other estimator and see what happens. One such estimator that deserves serious consideration is the least trimmed squares estimator with a breakdown point of .2 or perhaps .25. The Winsorized correlation estimator is another possibility, and some variations of this method, not covered here, might be better for general use. Certain members of the class of M-estimators, particularly the so-called adjusted M-estimator, might also be considered. Some commercial software contains several types of M-estimators, but care must be taken because not all of them have both a reasonably high breakdown point and a relatively small variance (standard error).

The second point that should be stressed has to do with extending regression estimators to situations where there is more than one predictor. All of the methods outlined here can be used with multiple predictors, but the breakdown point of the Theil-Sen estimator decreases as the number of predictors increases. With two predictors the breakdown point drops to about .13, a value that is dangerously low. An advantage of least trimmed squares regression is that the breakdown point does not decrease as the number of predictors increases. The same is true when using least trimmed absolute values and the adjusted M-estimator, details of which can be found in references given in the bibliographic notes. A definitive result on the breakdown point when using robust measures of correlation has not been derived. A guess is that when using a correlation with a high breakdown point, the breakdown point when estimating the slope is $1/(p + 1)$, where p is the number of predictors. So for the case of a single predictor, the breakdown point appears to be .5 when using correlations based on M-estimators, but this has not been formally established, and a speculation is that with multiple predictors, the breakdown point might be too low.

CORRELATION COEFFICIENTS BASED ON REGRESSION ESTIMATORS

Chapter 10 mentioned the problem of finding correlation coefficients that take into account the overall structure of the scatterplot of points being studied. Two such measures of correlation were mentioned in Section 10.9, but these measures are typically used to detect outliers; they are not used to test the hypothesis of independence or to summarize how well a particular regression line performs. The least trimmed squares regression line, the least absolute value regression line, and the deepest regression line could be used to define a measure of correlation that takes into account the overall structure of the points. One general possibility is to use

$$b_1 \frac{S(X)}{S(Y)},$$

where b_1 is some robust estimate of the slope and $S(X)$ and $S(Y)$ are some robust analog of the standard deviation, s. For example, one could choose b_1 to be the least trimmed squares estimate of the slope, and $S(X)$ and $S(Y)$ might be MAD or the Winsorized standard deviation. Arguments can be made for using measures of dispersion related to M-estimators, but this topic is best left for a more advanced book. Another possibility is to use some generalization based on the coefficient of determination. Both of these possibilities have received little attention so far.

A SUMMARY OF KEY POINTS

As a general rule, the breakdown point of any estimator should be at least .2. For simple regression, where there is a single predictor, the following methods satisfy this requirement:

- Theil-Sen
- least trimmed squares
- least trimmed absolute values
- the adjusted M-estimator
- methods based on robust correlation and variances for which a formal proof has not been published
- deepest regression line

All of these methods can be extended to the case where there are multiple predictors, but the breakdown point for Theil-Sen and the methods based on robust correlations decrease. In contrast, for the other methods just listed, the breakdown point is not affected. (The breakdown point of the deepest regression line is not affected, provided there is a linear association; otherwise it can be $1/(p+1)$, where again p indicates the number of predictors.)

As for ease of use, S-PLUS has a built-in function for doing least trimmed squares, but it uses the highest possible breakdown point. Libraries of S-PLUS functions are available for the estimators just listed, details of which can be found in Chapter 12.

Another practical consideration is execution time when using any estimator on a computer. All methods can be computed quickly except possibly the deepest regression line. Currently, versions of the deepest regression that can be used with S-PLUS are rather slow with a moderate sample size. (Fast FORTRAN programs are available from the web site described by Rousseeuw and van Driessen, 1999.) The technology for getting reasonably fast execution time using S-PLUS is available, but it has not been implemented.

As has been stressed, we want the variance of an estimator (its standard error) to be relatively small. Of particular concern are situations where there is heteroscedasticity. No single estimator always gives the best results, but based on this criterion, the following methods appear to perform relatively well:

- Theil-Sen

- least trimmed squares with a breakdown point of .2 or .25

- least trimmed absolute values with a breakdown point of .2 or .25

- the adjusted M-estimator

- methods based on robust correlation and variances

- deepest regression line

Generally, least trimmed squares beats least trimmed absolute value, but there are patterns of heteroscedasticity where the reverse is true and sometimes there is a practical advantage to setting the breakdown point to .5. Even when both X and Y are normal, there are situations where, due to heteroscedasticity, we get a smaller standard error using a breakdown point of .5 rather than .2 or .25. A guess is that for most situations, least trimmed squares generally beats least trimmed absolute values, but the empirical support for this view is rather weak. Another guess is that Theil-Sen typically has a smaller standard error than least trimmed squares or least trimmed absolute values, but situations arise where other robust methods have a smaller standard error than Theil-Sen. This illustrates why it is difficult—seemingly impossible—to recommend one estimator over all others.

A crude outline of how one might proceed is: Check for linearity using a smoother; a common strategy is to examine a plot of the residuals, but smoothers are more effective at detecting curvature. If fitting a straight line appears to be reasonable, use an estimator with a breakdown point between .2 and .3, and compare the results with an estimator that has a breakdown point of .5. If there is a large discrepancy, use plots or other diagnostic tools to determine why; details about such methods are covered in more advanced books. Otherwise, compute a confidence interval based on an estimator with a breakdown point of .2 or perhaps .25. One criticism of this relatively simple advice is that you might be in a situation where shorter confidence intervals are obtained using a breakdown point of .5.

Experience suggests that when there are two predictors, it is more common for curvature to be an issue than in situations where there is only one predictor. Again smoothers can help deal with this problem, but addressing curvature when there are three or more predictors remains a nontrivial problem; even with two predictors, advanced training can be required. It is easy to fit a flat plane to data. After reading your data into a computer, one command accomplishes this. But interesting associations might be revealed if advanced methods for dealing with curvature, which go well beyond the scope of this book, are used.

We conclude this chapter by stressing the following points:

- There is a collection of regression methods that have substantial advantages over more traditional techniques based on least squares.

- Even when there is one predictor, regression is a difficult problem that requires many tools to be done well.

- Despite our inability to find the one method that is best for general use, we can identify a strategy that is highly unacceptable: Use least squares and assume all is well.

- Even if a researcher picks a single method from the list of robust methods described in this chapter, one will generally get more accurate results on average. But to get the most out of data, multiple methods must be used.

BIBLIOGRAPHIC NOTES

For a more extensive list of robust regression methods and details on how to compute the least trimmed squares estimator, see Rousseeuw and Leroy

(1987). For the computational details associated with M-estimators and methods based on robust correlations and a description of relevant software, see Wilcox (1997). The adjusted M-estimator mentioned here is described in Section 8.5.4 of Wilcox (1997). For a recent discussion of least trimmed absolute value regression, see Hawkins and Olive (1999). For an approach to regression based on Kendall's tau, see Cliff (1996). (Also see Long, 1999.) For results supporting the use of the Theil-Sen estimator, see Wilcox (1998). Rousseeuw and Hubert (1999) describe regression depth.

CHAPTER *12*

ALTERNATE STRATEGIES

Space limitations prohibit a comprehensive discussion of the many statistical methods related to the techniques covered in previous chapters. However, a brief description of some of these alternate methods and comments about them should be made. One general issue is the role of so-called *ranked-based* or *nonparametric methods* for comparing groups and performing regression. Such methods are commonly recommended for dealing with nonnormality, so it is important to compare and contrast them with methods based on robust measures of location. Permutation tests are yet another method sometimes recommended for dealing with nonnormality. Another important issue is extending methods for comparing groups to more complicated experimental designs. Finally, some comments about software will be made.

RANKED-BASED METHODS FOR COMPARING TWO GROUPS

We begin by expanding on the problem in Chapter 9 where the goal was to compare two independent groups of subjects. For example, Table 12.1 reports data from a study on the effects of ozone on weight gain in rats. The experimental group consisted of twenty-two seventy-day-old rats kept in an ozone environment for seven days. A control group of twenty-three rats the same age was kept in an ozone-free environment. An issue is deciding whether ozone affects weight gain, and if so, characterizing how much. One way of comparing these two groups is in terms of the typical amount of weight gained in the control group compared to the experimental group. The common approach would be to compare means, or one could compare some robust measure of location such as a 20 percent trimmed mean or the one-step M-estimator, as described in Chapter 9. There is an alternative strategy that might be considered. To describe it, let p be the probability that a randomly sampled observation from the first group is less than a randomly sampled observation from the second. In the illustration, p is the probability that a randomly sampled rat assigned to the control group would gain less weight than a randomly sampled rat assigned to the ozone group. If in terms of weight gained it makes absolutely no difference to which group a rat is assigned, and in fact the probability curves for both groups are identical, then $p = 1/2$. Moreover, a reasonable method for characterizing how the two groups differ is in terms of p. If $p = .8$, for example, this provides some sense of how much the first group differs from the second. In terms of establishing that the groups differ, one could test $H_0 : p = 1/2$, the hypothesis that p is exactly equal to .5.

TABLE 12.1 • WEIGHT GAIN OF RATS IN OZONE EXPERIMENT

Control	41.0	38.4	24.4	25.9	21.9	18.3	13.1	27.3	28.5	−16.9
Ozone	10.1	6.1	20.4	7.3	14.3	15.5	−9.9	6.8	28.2	17.9
Control	26.0	17.4	21.8	15.4	27.4	19.2	22.4	17.7	26.0	29.4
Ozone	−9.0	−12.9	14.0	6.6	12.1	15.7	39.9	−15.9	54.6	−14.7
Control	21.4	26.6	22.7							
Ozone	44.1	−9.0								

Estimating p can be accomplished as follows. Consider the first observation in the second (ozone) group, which is 10.1. It can be seen that only one value in the first group is less than 10.1. Let's denote this by $V_1 = 1$. If three of the observations in the first group were less than the first observation in the second, then we would denote this by $V_1 = 3$. Notice that V_1/n_1 estimates p, where n_1 is the number of observations in the first group. In the illustration, $V_1/n_1 = 1/23$, meaning that the proportion of observations in the first group that are less than 10.1 is 1/23. The second observation in the second group is 6.1, and exactly one observation in the first group is less than this value, which we denote by $V_2 = 1$. Again our estimate of p is $1/23$. The third observation in the second group is 20.4; it exceeds seven of the observations in the first group. We write this as $V_3 = 7$, and now our estimate of p is $7/23$. Of course, we can continue in this manner for all of the observations in the second group, yielding n_2 estimates of p, where n_2 is the number of observations in the second group. A natural way to combine all these estimates of p is to average them, and this is what is done in practice. In more formal terms, the estimate of p is

$$\hat{p} = \frac{1}{n_1 n_2}(V_1 + \cdots + V_n).$$

There is a classic method for making inferences about p based on $\hat{p}$. Called the *Wilcoxon test*, which is also known as the *Mann-Whitney U test*, it is possible to get exact control over the probability of a Type I error if, in addition to random sampling, the two groups being compared have identical probability curves. In particular, the two groups have equal variances. Tables for implementing the method have been constructed for situations where the sample sizes are small. For moderate to large sample sizes, an approximate method is typically used and is based on the central limit theorem. Letting

$$\sigma_u = \sqrt{\left(\frac{n_1 n_2(n_1 + n_2 + 1)}{12}\right)},$$

the test statistic typically used when making inferences about p is

$$Z = \frac{\hat{p} - .5}{\sigma_u/(n_1 n_2)}. \tag{12.1}$$

When the groups being compared have identical distributions, $\sigma_u/(n_1 n_2)$ estimates the standard error of $\hat{p}$ and Z has, approximately, a standard normal probability curve. The hypothesis that $p = 1/2$ is rejected if Z is sufficiently large or small. Details and illustrations can be found in most introductory statistics books.

It should be pointed out that some theoretically oriented books describe the Wilcoxon-Mann-Whitney test as a test of the hypothesis that two groups of individuals have identical probability curves rather than a test designed to make inferences about p. When we say that this test provides exact control over the probability of a Type I error, this refers to the hypothesis that the probability curves are identical. Notice that the numerator of Z, given by equation 12.1, is based on $\hat{p} - .5$. So in the particular case where the true probability p is .5, on average the numerator of Z will be zero. (In formal terms, the expected value of $\hat{p} - .5$ is $E(\hat{p} - .5) = 0$.) This suggests that the Wilcoxon-Mann-Whitney method is satisfactory for making inferences about p, but there is a feature of the test that should be kept in mind. The estimate of the standard error of $\hat{p}$, $\sigma_u/(n_1 n_2)$, is derived under the assumption that the two groups have identical probability curves. If the probability curves differ, this estimate of the standard error is incorrect.

To illustrate one implication of this result, consider two normal probability curves that both have a mean of zero; the first has a standard deviation of one and the other has a standard deviation of ten. Then the probability that a randomly sampled observation from the first group is less than a randomly sampled observation from the second is $p = .5$. So we should not reject the hypothesis that $p = .5$. But if we sample forty observations from each group and apply the Wilcoxon-Mann-Whitney test at the .05 level, the actual probability of rejecting is approximately .09. So if our goal is to control the probability of a Type I error when making inferences about p, the Wilcoxon-Mann-Whitney test can be unsatisfactory. In some sense this is not surprising because the two probability curves differ and the Wilcoxon-Mann-Whitney test is based on the assumption that the curves are exactly the same. But from another point of view this result is unexpected because the test is based on an estimate of how much p differs from .5, and in the situation under consideration it does not differ at all. Insofar as we want to make inferences about p, this is unacceptable. In addition, there are problems when computing a confidence interval for p because the wrong standard error is being used when probability curves have unequal variances. Unequal variances can also result in poor power properties.

Notice the similarity between the properties of Student's T test, described in Chapter 5, and the properties of the Wilcoxon-Mann-Whitney test. Both tests are designed to be sensitive to a particular feature of how the groups differ. The two-sample Student's T is intended to be sensitive to differences between the population means, and the Wilcoxon-Mann-Whitney test is designed to be sensitive to situations where p differs from .5. But in reality, both methods test the hypothesis that distributions are identical, and both are sensitive to a myriad of ways in which the probability curves might differ.

In recent years there has been interest in finding inferential methods for p that remain accurate when groups have unequal variances and more generally when the probability curves differ in shape. Significant progress has been made, but another practical problem arises: How should one deal with tied values? For the first rat in Table 12.1, the weight gain was 41 grams. If another rat had gained 41 grams, we say that there are tied values (meaning that a particular value occurs more than once). When using a rank-based method that allows unequal variances, special methods are required to handle the common situation of tied values. Here we merely note that successful methods have been derived and are summarized in a book by N. Cliff. What is more important here is understanding the relative merits of using these rank-based methods compared to the techniques outlined in Chapter 9.

Although it is not obvious from the description of the Wilcoxon-Mann-Whitney test given here, it is well known that it and its heteroscedastic analogs can be reformulated in terms of ranks. As noted in Chapter 10, the smallest observation among a batch of numbers gets a rank of 1, and if we lower the smallest value, its rank does not change. In a similar manner, if we increase the largest value, its rank does not change. This suggests that the probability of detecting situations where p differs from .5 (power) can remain relatively high when sampling from probability curves that are likely to produce outliers, and this speculation is correct.

So, we have two strategies for comparing groups that give good control over the probability of a Type I error, and they have good power properties when sampling from nonnormal probability curves: methods based on robust measures of location and recently developed methods based on ranks. How do we choose which approach to use? In terms of maximizing our chances of detecting a true difference, there is no clear-cut choice. Situations can be constructed where comparing trimmed means or M-estimators will mean substantially more power, but situations can be constructed where the reverse is true. Experience with data is not much help either. Sometimes we reject when using measures of location, but not when making inferences about p, and we encounter situations where the opposite happens. For example, if we compare the two groups in Table 12.1 using the method in Cliff's book, the estimate of p is .52 and a .95 confidence interval for p is (.14, .77). Based on p there is little difference between the methods, and because the confidence interval contains the value .5, we are unable to reject the hypothesis that $p = .5$. That is, there is no empirical evidence that weight gains differ for rats in an ozone environment compared to rats in an ozonefree environment. But if we compare 20 percent trimmed means, we reject the hypothesis that the population trimmed means are equal using the percentile t bootstrap method. For the first group the sample trimmed mean is 23.3, for the second it is 9.2, and a .95 confidence interval for the difference between the population trimmed means is (3.8, 21.4). Again, this merely illustrates that different perspectives can yield different conclusions about whether groups differ and by how much. Differences between trimmed means, for example, can give little or no indication about what the magnitude of p might be, and p does not necessarily tell us anything about the difference between trimmed means. For similar reasons, the Wilcoxon-Mann-Whitney test is unsatisfactory when trying to make inferences about medians (e.g., Kendall and Stuart, 1973; Hettmansperger, 1984).

Figure 12.1 illustrates one reason why rank-based methods might give different results from some robust measure of location. The two probability curves have equal trimmed means, so the hypothesis of equal trimmed means should not be rejected. But $p = .42$, so we should reject the hypothesis that $p = .5$, and the probability of rejecting approaches one when using Cliff's method, as the sample sizes get large. But we can shift the symmetric curve in Figure 12.1 so that $p = .5$ and the trimmed means are not equal. So now the probability of rejecting when comparing trimmed means goes to one as the sample sizes get large, but when using Cliff's method the probability of rejecting goes to .05 when testing $H_0 : p = .5$ at the .05 level.

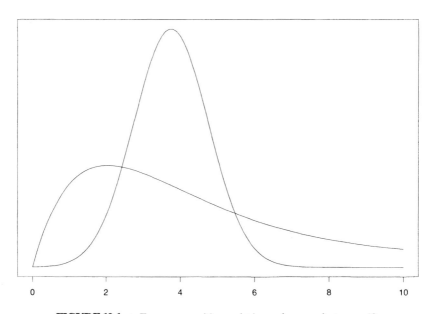

FIGURE 12.1 • Two curves with equal trimmed means, but $p = .42$.

When comparing multiple groups of individuals, again there are standard rank-based methods that are typically covered in an introductory statistics course that should be replaced by more modern techniques. The computational details surrounding modern methods are complex and difficult to explain in a straightforward manner. But details and software are available, as will be pointed out at the end of this chapter.

PERMUTATION TESTS

There are many computer-intensive methods beyond the bootstrap techniques covered in Chapter 6. For completeness, we briefly describe one more method for comparing two (independent) groups because it is sometimes recommended when distributions are nonnormal. The idea was first introduced by R. A. Fisher in the 1930's. Called a *permutation test*, it is designed to test the hypothesis that two groups have identical probability curves. There is in fact a large class of methods among permutation tests, but for brevity attention is focused on a version based on the sample means.

To illustrate the basic idea, imagine that the sample means are $\bar{X}_1 = 26$ and $\bar{X}_2 = 18$, in which case the estimated difference between the population means is $\bar{X}_1 - \bar{X}_2 = 26 - 18 = 8$. As in previous chapters, let n_1 and n_2 be the sample sizes corresponding to each group. The permutation test is designed to determine whether the observed difference between the sample means is large enough to reject the hypothesis that the two groups have identical probability curves. The strategy is similar to the bootstrap covered in Chapter 9, but it differs in two important ways. First, rather than resampling n_1 observations from the first group and then resampling n_2 observations from the second, the observations are first pooled. From these pooled values, we resample n_1 observations *without* replacement. This is the second major difference from the bootstrap method, where resampling is done with replacement. The sample mean for these n_1 observations is computed, then one computes the sample mean for the remaining n_2 observations, and the difference between the resulting sample means is recorded. This process is repeated many times, and if the middle 95 percent of the resulting differences do not contain $\bar{X}_1 - \bar{X}_2$ (which is 8 in the illustration), reject the hypothesis of identical probability curves at the .05 level.

Explained another way, if the hypothesis of identical distributions is correct, then any observation we make could have come from either group being compared. The strategy is to combine the observations and randomly choose n_1 observations and temporarily assume they came from the first group. The remaining n_2 observations are temporarily assumed to have come from the second group. Said yet another way, we permute the $n_1 + n_2$ observations in a random and arbitrary way and temporarily assume the first n_1 observations came from the first group. Then we compute the difference between the group means and repeat this process many times. If the null hypothesis is true and we want the probability of a Type I error to be .05, then we reject the null hypothesis if the middle 95 percent of the differences just computed does not contain $\bar{X}_1 - \bar{X}_2$, the difference between the sample means based on the original data.

A positive feature of the permutation test is that it provides exact control over the probability of a Type I error when testing the hypothesis of identical distributions. A negative feature is that it does not yield confidence intervals and generally fails to tell us how the distributions differ and by how much. For example, although the version of the permutation test just given is based on the sample mean, it is sensitive to other features of the distributions being compared. For example, Boik (1987) demonstrates that even when sampling from normal distributions with equal means, the probability of rejecting is affected by the degree to which the variances differ. In practical terms, if you reject with the permutation test just described, it is reasonable to conclude that the distributions differ. But if the main goal is to determine how and to what extent the population means differ, the permutation test is unsatisfactory. More generally, it fails to indicate how distributions differ in terms of any of the measures of location and scale covered in this book.

EXTENSION TO OTHER EXPERIMENTAL DESIGNS

The goal is often not to compare just two groups of individuals, but three or more groups. For example, clinical psychologists have long tried to understand schizophrenia. One issue of interest to some researchers is whether different groups of subjects differ in terms of measures of skin resistance. In such a study by S. Mednick, four groups of individuals were identified: (1) no schizophrenic spectrum disorder, (2) schizotypal or paranoid personality disorder, (3) schizophrenia, predominantly negative symptoms, and (4) schizophrenia, predominantly positive symptoms. Table 12.2 presents the first ten observations for each group, where the entries are measures of skin resistance (in Ohms) following presentation of a generalization stimulus. (The actual sample sizes used in the study were larger; only the first ten observations for each group are listed here.) You could, of course, compare all pairs of groups using methods described in Chapter 9. That is, you could compare the first group to the second, then compare the first group to the third, and so on. But one concern is that as the number of comparisons increases, the probability of at least one Type I error will also increase. There are many methods aimed at dealing with this issue. A detailed description about such methods goes well beyond the scope of this book, but a few comments about modern trends and developments might be helpful.

TABLE 12.2 • MEASURES OF SKIN RESISTANCE FOR FOUR GROUPS OF SUBJECTS

(NO SCHIZ.)	(SCHIZOTYPAL)	(SCHIZ. NEG.)	(SCHIZ. POS.)
0.49959	0.24792	0.25089	0.37667
0.23457	0.00000	0.00000	0.43561
0.26505	0.00000	0.00000	0.72968
0.27910	0.39062	0.00000	0.26285
0.00000	0.34841	0.11459	0.22526
0.00000	0.00000	0.79480	0.34903
0.00000	0.20690	0.17655	0.24482
0.14109	0.44428	0.00000	0.41096
0.00000	0.00000	0.15860	0.08679
1.34099	0.31802	0.00000	0.87532

First, even if one were to restrict attention to comparing means, despite the many problems that have been described, there have been major advances and improvements over conventional homoscedastic methods. It might be hoped that the problems with Student's T method for comparing groups, described in Chapter 9, diminish as we move toward situations where we compare multiple groups, but the exact opposite happens. Second, all of the conventional homoscedastic methods have heteroscedastic analogs that can be used with trimmed means. In most cases, M-estimators can also be used, but there are some situations where M-estimators are less convenient than trimmed means from a technical point of view. A textbook by Maxwell and Delaney, published in 1990, covers some of the more modern methods for comparing means, but there has been substantial progress since then, including bootstrap methods that offer more accurate inferences for reasons outlined in Chapter 6. Rank-based (or nonparametric) methods remain a viable option, but again, modern heteroscedastic methods are recommended. The main point here is that books and convenient software describing these methods are available.

In some situations it is desirable to compare dependent groups. For example, in a study on the effectiveness of a treatment for depression, part of the study might be aimed at assessing how measures of depression change before and after the treatment. Because the measures are taken on the same individual at two different times, it is unreasonable to assume that the measures at time 1 are independent of those taken at time 2. Here it is merely noted that methods for dealing with this problem have been derived for when working with robust measures of location, details of which can be found in books listed in the bibliographic notes at the end of this chapter.

COMMENTS ON THE FAMILYWISE ERROR RATE

Again consider the data in Table 12.2 and imagine that you want to compare all pairs of groups. The goal is to compare group 1 to group 2, group 1 to group 3, and so on. There are six pairs of groups to be compared. Of course, if for every pair of groups we compare the trimmed means, for example, there is the probability of making a Type I error—declaring a difference when in fact no difference exists. So among the family of all six comparisons, there is some probability of making one or more Type I errors. The *familywise error rate* (FWE) is the probability of at least one Type I error among the entire family of comparisons to be made. How can we control FWE? If, for example, we compare each group with the probability of a Type I error set at .05, the FWE will be greater than .05 by an amount that is not easily determined. If we want the FWE to be .05, how can we adjust the individual comparisons to accomplish this goal?

There is a vast literature on this topic and book-length descriptions of the strategies that have been proposed. All of the methods developed explicitly for means can be extended to trimmed means, and software for applying some of these methods can be found in Wilcox (1997). For small to moderately large sample sizes, bootstrap techniques again appear to have practical value and are recommended based on what is currently known. Although the details about such techniques go beyond the scope of this book, a few comments might be useful.

Consider the problem of comparing means. One general approach to controlling FWE is to begin by testing the hypothesis that all the groups have a common mean. For the data in Table 12.1, the goal is to test

$$H_0 : \mu_1 = \mu_2 = \mu_3 = \mu_4, \qquad (12.2)$$

where μ_1, μ_2, μ_3, and μ_4 are the population means corresponding to the four groups. If a nonsignificant result is obtained, meaning you fail to reject the hypothesis that all of the population means are equal, the analysis stops and you fail to find any differences between any two groups under consideration. If you do reject, then pairs of groups are compared using one of many methods that have been proposed, but no details are given here. What is important is the following consideration. Imagine that the first three groups have a normal probability curve with equal variances and the means differ to the point that power is high. That is, if we were to compare these three groups, ignoring the fourth group, there is a high probability of rejecting the hypothesis of equal means. But now imagine that the probability curve for the fourth group is a mixed normal, which was described in Chapter 7. This can drastically lower power. In fact, a very small departure from normality in the fourth group can make it highly unlikely that the differences among the first three groups will be detected. This problem can be addressed by comparing trimmed means instead.

There are also methods for controlling FWE where one does not begin with testing Equation 12.1. Rather, for each pair of groups, one of the methods described in Chapter 9 is applied, but the individual Type I error probabilities are adjusted so that FWE does not exceed some specified amount. There are advantages and disadvantages associated with these techniques, and the better methods can be used with robust measures of location. Included are techniques for comparing dependent groups, and again certain types of bootstrap methods appear to be best when sample sizes are small. Some of the most recent methods are described in Wilcox (1997), but it appears that even better methods will soon be available.

To emphasize an important point, the data in Table 12.2 illustrate that the choice of method can be crucial. If all pairwise comparisons of the 20 percent trimmed means are compared with the S-PLUS function lincon, described in Wilcox (1997), or with mcppb, which is mentioned on page 241, it is concluded that groups 3 and 4 differ with FWE set at .05. If we compare the groups based on p and control FWE with the function cidmul (described later), or if we set the trimming to zero so that means are being compared, we no longer reject.

RANK-BASED METHODS FOR COMPARING MULTIPLE GROUPS

It is briefly mentioned that rank-based methods are available for comparing multiple groups that offer potentially useful alternatives to methods based on robust measures of location. There are standard methods typically covered in an introductory course: The Kruskal-Wallis test compares independent groups, and the Friedman test compares dependent groups. These methods appear to perform very well in terms of Type I errors when comparing groups having completely identical distributions. But when groups differ, they become unsatisfactory for reasons similar to why the Wilcoxon-Mann-Whitney is unsatisfactory. Various methods have been proposed for dealing with these problems, summaries of which can be found in Wilcox (1997), and some additional methods are covered in Cliff (1996).

REGRESSION BASED ON RANKED RESIDUALS

A variety of methods for fitting a straight line to data have been proposed where, instead of choosing the slope and intercept as described in Chapter 11, the rank of the residuals is used instead. That is, given a choice for the slope and intercept, assign the value 1 to the smallest residual, a value of 2 to the second smallest, and so on. Then the performance of a particular regression line is judged based on some function of the resulting ranks. The goal, then, is to find the regression line that minimizes this function. Several reasonable functions have been proposed and are covered in more advanced books. For some of these methods, excellent control over the probability of a Type I error can be had, even when there is heteroscedasticity. A possible concern is that power, the probability of detecting situations where the slope does indeed differ from zero, can be poor relative to other methods covered in Chapter 11. However, in fairness, situations arise where some of these methods offer a substantial advantage over competing techniques. For example, if X has a normal distribution and Y, given X, has a nonnormal distribution, a method proposed by J. Naranjo and T. Hettmansperger in 1994 offers an advantage over least squares, in some cases by a substantial amount. Moreover, their method competes well with other robust regression methods. Unfortunately, when X has a skewed probability curve, situations arise where it performs very poorly relative to other approaches covered here. So this is yet another approach to regression that might have practical value in a given situation, but routine use of this approach, to the exclusion of all other methods one might consider, cannot be recommended.

SOFTWARE

Naturally, there is always a lag between theoretical-methodological developments in statistics and software. Software developed specifically for employing the bootstrap is now available, but what is needed is easy-to-use software that provides access to the methods described in Chapters 8 through 12 of this book. Such software is available, but not commercially, meaning that free software, written by researchers, must be used.

Vargha (1999) has supplied Ministat, which is very easy to use. There is also a library of S-PLUS functions written by the author and available via anonymous ftp at ftp.usc.edu.[1] Once connected, change directories to pub/wilcox, then download the file allfun. The file update_info, located in the same directory, contains information on functions recently updated or added. Versions of these functions could be written in SAS, but this has not been done yet.

Details on how to use the bulk of these S-PLUS functions can be found in Wilcox (1997). Included are some modern rank-based methods. For convenience, a few of the newer functions, which are not described in Wilcox (1997), are listed here.

COMPARING GROUPS

mcppb: Performs all pairwise comparisons of multiple independent groups using trimmed means and a percentile bootstrap method. When comparing two groups only, with 20 percent trimming, it appears that there is some advantage to using the percentile bootstrap rather than the percentile t. When comparing multiple groups, the percentile method has a distinct advantage (Wilcox, in press b). However, when comparing means, this function is not recommended.

rmmcp: Same as mcppb, but dependent groups are compared.

cid: Performs Cliff's heteroscedastic analog of the Wilcoxon-Mann-Whitney test when there are tied values.

cidmul: An extension of Cliff's method for comparing all pairs of independent groups that controls the probability of at least one Type I error.

---

[1] S-PLUS is a product of MathSoft, Inc.

CORRELATION AND REGRESSION

tsreg: Computes the Theil-Sen estimator when there is only one predictor.

tsgreg: Computes the Theil-Sen estimator when there is more than one predictor.

lts1reg: Computes the least trimmed squares estimator when there is only one predictor. (S-PLUS has a built-in function, ltsreg, that performs least trimmed squares with the breakdown point set to .5.)

ltsgreg: Computes the least trimmed squares estimator when there is more than one predictor.

ltareg: Computes the least trimmed absolute value estimator when there is only one predictor.

ltagreg: Computes the least trimmed squares estimator when there is more than one predictor.

depreg: Computes the deepest regression line when there is only one predictor.

pcorb: A bootstrap method for computing a confidence interval for Pearson's correlation that allows heteroscedasticity.

corb: A bootstrap method for computing a confidence interval for any of the robust correlations covered in this book. The method allows heteroscedasticity and can be used with other robust correlations described in Wilcox (1997).

indt: A method for detecting dependence that is sensitive to various types of associations that might be missed when using Pearson's correlation or some of its robust analogs. (See Wilcox, in press.)

A SUMMARY OF KEY POINTS

- Rank-based methods provide an alternate method for dealing with low power due to outliers. The better-known methods, such as the Wilcoxon-Mann-Whitney test, the Kruskal-Wallis test, and Friedman's test are known to have practical problems when distributions differ. It is recommended that they be replaced by the techniques summarized in Wilcox (1997) and Cliff (1996).

- The choice between modern methods based on robust measures of location or modern (heteroscedastic) rank-based methods is not straightforward. The two approaches are sensitive to different features of the data. That is, they provide different ways of characterizing how groups differ. Consequently, in terms of power, rank-based methods provide an advantage in some situations but not others. In exploratory studies, perhaps both should be used.

- Permutation tests are useful when testing the hypothesis that distributions are identical and exact control over the probability of a Type I error is desired, but they are not very useful when the goal is to determine how groups differ.

- Methods based on robust measures of location and heteroscedastic rank-based methods can be extended to situations where the goal is to compare multiple groups. Included are methods for studies having multiple factors (two-way and three-way designs, repeated measures, and split-plot designs). The technical problems associated with rank-based methods are particularly difficult, but practical methods have been derived and easy-to-use software is available.

- Yet another approach to regression is to judge the fit of a line to a scatterplot of points based on the ranks of the residuals. Several explicit possibilities have been proposed and studied. Although such methods can be relatively unsatisfactory in some situations, other situations arise where they compare well to competing techniques, so they seem worth considering in applied work.

BIBLIOGRAPHIC NOTES

For books dedicated to controlling FWE when comparing means, see Hochberg and Tamhane (1987) or Hsu (1996). Westfall and Young (1993) describe bootstrap methods for controlling FWE when comparing means. These bootstrap methods are readily extended to robust measures of location (Wilcox, 1997). For a recent description of various rank-based approaches to regression, see Hettmansperger and McKean (1998). For details about the rank-based regression method mentioned in Section 12.3, see Naranjo and Hettmansperger (1994). For heteroscedastic rank-based methods for comparing groups, see Cliff (1996). For some extensions of rank-based methods for comparing multiple groups not covered by Cliff, see Wilcox (1997).

APPENDIX A

This appendix summarizes some basic principles related to summation and expected values.

Basic summation notation can be described as follows. We have n observations, which we label $X_1, X_2, \ldots, X_n$. Then

$$\sum X_i = X_1 + X_2 + \cdots + X_n,$$

where X_i represents the ith observation, $i = 1, \ldots, n$. For example, if we measure Mary, Fred, and Barbara and get the values 6, 12, and 10, respectively, then $n = 3$ and $\sum X_i = 6 + 12 + 10 = 28$. The sample mean is

$$\bar{X} = \frac{1}{n} \sum X_i.$$

Let $p(x)$ be the probability of observing the value x. The expected value of some measure X is

$$E(X) = \sum x p(x),$$

where $\sum$ indicates summation over all possible values of x. The population mean is $\mu = E(X)$. For example, imagine that individuals rate the nutritional benefits of a food on a four-point scale: 1, 2, 3, and 4 and that for the population of adults the corresponding probabilities are .2, .3, .4, and .1. So when an adult is interviewed, the probability of a rating 1 is $p(1) = .2$. The expected or average rating for all adults is

$$1p(1) + 2p(2) + 3p(3) + 4p(4) = 2.4.$$

That is, the population mean is 2.4.

Two measures are said to be *identically distributed* if the possible outcomes and corresponding probabilities are identical. So in the last illustration, if we select an adult and the probabilities corresponding to 1, 2, 3, and 4 are .2, .3, .4, and .1 and the same is true for the second adult we pick, then the two measures are said to be identically distributed. In symbols, X_1 and X_2 are identically distributed if both have the same possible outcomes and corresponding probabilities. If the observations $X_1, \ldots, X_n$ are independent and identically distributed, they are said to be a *random sample* of size n.

The population variance associated with a single observation (X) is

$$\sigma^2 = E[(X - \mu)^2],$$

the average squared distance of an observation from the population mean. For the nutritional rating example, the population variance is

$$\sigma^2 = (1 - 2.4)^2(.2) + (2 - 2.4)^2(.3) + (3 - 2.4)^2(.4) + 4 - 2.4)^2(.1) = .84.$$

It can be shown that for any constant c,

$$E(cX) = cE(X) = c\mu.$$

Furthermore, letting VAR(X) indicate the variance of X,

$$\text{VAR}(cX) = c^2 \text{VAR}(X) = c^2\sigma^2.$$

Also,

$$\text{VAR}(X_1 + X_2) = \text{VAR}(X_1) + \text{VAR}(X_2) + 2\rho\sqrt{\text{VAR}(X_1)\text{VAR}(X_2)},$$

where ρ is the population correlation between X_1 and X_2.

From these rules of expected values it can be seen that if $X_1, \ldots, X_n$ is a random sample, in which case every pair of observations is independent and has $\rho = 0$, then

$$\text{VAR}(\bar{X}) = \frac{1}{n^2}(\sigma^2 + \cdots + \sigma^2) = \sigma^2/n.$$

APPENDIX A   247

That is, the variance (or squared standard error) of the sample mean is the variance of a single observation divided by the sample size. This fundamental result forms the basis of Laplace's confidence interval for the population mean.

Let $X_{(1)} \leq X_{(2)} \leq \cdots \leq X_{(n)}$ be the n observations written in ascending order. If we remove the g smallest and g largest and average the remaining observations, we get a trimmed mean. We saw, however, that the remaining observations are dependent. In fact, their correlations are not equal to zero. So the derivation of the variance of the sample mean does not generalize to the problem of determining the variance of the trimmed mean. In practical terms, methods based on means do not generalize to situations where observations are trimmed because this results in using the wrong standard error.

REFERENCES

Barnett, V., & Lewis, T. (1994). *Outliers in Statistical Data*. New York: Wiley.

Berger, R. L., & Hsu, J. C. (1996). Bioequivalence trials, intersection-union tests and equivalence confidence sets. *Statistical Science, 11*, 283–319.

Boik, R. J. (1987). The Fisher-Pitman permutation test: A non-robust alternative to the normal theory F test when variances are heterogeneous. *British Journal of Mathematical and Statistical Psychology, 40*, 26–42.

Box, G. E. P. (1954). Some theorems on quadratic forms applied in the study of analysis of variance problems, I. Effect of inequality of variance in the one-way model. *Annals of Mathematical Statistics, 25*, 290–302.

Bradley, J. V. (1978) Robustness? *British Journal of Mathematical and Statistical Psychology, 31*, 144–52.

Cliff, N. (1996). *Ordinal Methods for Behavioral Data Analysis*. Mahwah, NJ: Erlbaum.

Cohen, J. (1977). *Statistical Power Analysis for the Behavioral Sciences*. New York: Academic Press.

Cressie, N. A. C., & Whitford, H. J. (1986). How to use the two sample t-test. *Biometrical Journal, 28*, 131–48.

Davison, A. C., & Hinkley, D. V. (1997). *Bootstrap Methods and Their Application.* Cambridge: Cambridge University Press.

Efron, B., & Tibshirani, R. J. (1993). *An Introduction to the Bootstrap.* New York: Chapman & Hall.

Ellis, R. L. (1844). On the method of least squares. *Transactions of the Cambridge Philosophical Society 8,* 204–19.

Freedman, D., & Diaconis, P. (1982). On inconsistent M-estimators. *Annals of Statistics, 10,* 454–61.

Glass, G. V., Peckham, P. D., & Sanders, J. R. (1972). Consequences of failure to meet assumptions underlying the analysis of variance and covariance. *Review of Educational Research, 42,* 237–88.

Goldberg, K. M., & Iglewicz, B. (1992). Bivariate extensions of the boxplot. *Technometrics, 34,* 307–20.

Hald, A. (1998). *A History of Mathematical Statistics.* New York: Wiley.

Hall, P. (1986). On the number of bootstrap simulations required to construct a confidence interval. *Annals of Statistics, 14,* 1431–52.

Hampel, F. R., Ronchetti, E. M., Rousseeuw, P. J., & Stahel, W. A. (1986). *Robust Statistics.* New York: Wiley.

Hastie, T. J., & Tibshirani, R. J. (1990). *Generalized Additive Models.* New York: Chapman & Hall.

Hawkins, D. M., & Olive, D. (1999). Applications and algorithm for least trimmed sum of absolute deviations regression. *Computational Statistics & Data Analysis, 32,* 119–34.

Hettmansperger, T. P. (1984). *Statistical Inference Based on Ranks.* New York: Wiley.

Hettmansperger, T. P., & McKean, J. W. (1998). *Robust Nonparametric Statistical Methods.* London: Arnold.

Hochberg, Y., & Tamhane, A. C (1987). *Multiple Comparison Procedures.* New York: Wiley.

Hsu, J. C. (1996). *Multiple Comparisons.* London: Chapman and Hall.

Huber, P. (1993). Projection pursuit and robustness. In S. Morgenthaler, E. Ronchetti, & W. Stahel (eds.) *New Directions in Statistical Data Analysis and Robustness,* pp. 139–46. Boston: Birkhäuser Verlag.

Huber, P. J. (1964). Robust estimation of location. *Annals of Mathematical Statistics, 35,* 73–101.

Huber, P. J. (1981). *Robust Statistics.* New York: Wiley.

Kendall, M. G., & Stuart, A. (1973). *The Advanced Theory of Statistics,* Vol. 2. New York: Hafner.

Keselman, H. J., Huberty, C. J., Lix, L. M., Olejnik, S., Cribbie, R. A., Donohue, B., Kowalchuk, R. K., Lowman, L. L., Petosky, M. D., Keselman, J. C., & Levin, J. R. (1998). Statistical practices of educational researchers: An analysis of their ANOVA, MANOVA, and ANCOVA analyses. *Review of Educational Research, 68,* 350–86.

Liu, R. Y., Parelius, J. M., & Singh, K. (1999). Multivariate analysis by data depth: Descriptive statistics. *Annals of Statistics, 27*, 783–840.

Long, J. D. (1999). A confidence interval for ordinal multiple regression weights. *Psychological Methods, 4*, 315–30.

Luh, W.-M., & Guo, J.-H. (1999). A powerful transformation trimmed mean method for one-way fixed effects ANOVA model under non-normality and inequality of variances. *British Journal of Mathematical and Statistical Psychology, 52*, 303–20.

Maxwell, S. E., & Delaney, H. D. (1990). *Designing Experiments and Analyzing Data: A Model Comparison Perspective*. Belmont, CA: Wadsworth.

Naranjo, J. D., & Hettmansperger, T. P. (1994). Bounded-influence rank regression. *Journal of the Royal Statistical Society, Series B, 56*, 209–20.

Ramsey, P. H. (1980). Exact type I error rates for robustness of Student's T test with unequal variances. *Journal of Educational Statistics, 5*, 337–49.

Rasmussen, J. L. (1989). Data transformation, Type I error rate and power. *British Journal of Mathematical and Statistical Psychology, 42*, 203–11.

Rosenberger, J. L., & Gasko, M. (1983). Comparing location estimators: Trimmed means, medians, and trimean. In D. C. Hoaglin, F. Mosteller, and J. W. Tukey, (eds.) *Understanding Robust and Exploratory Data Analysis*. New York: Wiley.

Rousseeuw, P. J. (1984). Least median of squares regression. *Journal of the American Statistical Association, 79*, 871–80.

Rousseeuw, P. J., & Leroy, A. M. (1987). *Robust Regression and Outlier Detection*. New York: Wiley.

Rousseeuw, P. J., & van Driessen, K. (1999). A fast algorithm for the minimum covariance determinant estimator. *Technometrics, 41*, 212–23.

Rousseeuw, P. J., & van Zomeren, B. C. (1990). Unmasking multivariate outliers and leverage points. *Journal of the American Statistical Association, 85*, 633–9.

Rousseeuw, P. J., & Hubert, M. (1999). Regression depth. *Journal of the American Statistical Association, 94*, 388–402.

Rousseeuw, P. J., Ruts, I., & Tukey, J. W. (1999). The bagplot: A bivariate boxplot. *American Statistician, 53*, 382–7.

Shao, J., & Tu, D. (1995). *The Jackknife and the Bootstrap*. New York: Springer-Verlag.

Sockett, E. B., Daneman, D., Carlson, C., & Ehrich, R. M. (1987). Factors affecting and patterns of residual insulin secretion during the first year of type I (insulin dependent) diabetes mellitus in children. *Diabetes, 30*, 453-9.

Staudte, R. G., & Sheather, S. J. (1990). *Robust Estimation and Testing*. New York: Wiley.

Tukey, J. W. (1960). A survey of sampling from contaminated normal distributions. In I. Olkin et al. (eds.) *Contributions to Probability and Statistics*. Stanford, CA: Stanford University Press.

Tukey, J. W. (1977). *Exploratory Data Analysis*. Reading, MA: Addison-Wesley.
Vargha, A. (1999). Ministat, Version 3.1 Manual. Budapest: Polya Publisher.
Westfall, P. H., & Young, S. S. (1993). *Resampling Based Multiple Testing*. New York: Wiley.
Wilcox, R. R. (1996a). Confidence intervals for the slope of a regression line when the error term has non-constant variance. *Computational Statistics & Data Analysis*, 22, 89–98.
Wilcox, R. R. (1996b). A note on testing hypotheses about trimmed means. *Biometrical Journal*, 38, 173–80.
Wilcox, R. R. (1997). *Introduction to Robust Estimation and Hypothesis Testing*. San Diego: Academic Press.
Wilcox, R. R. (1998). A note on the Theil-Sen regression estimator when the regressor is random and the error term is heteroscedastic. *Biometrical Journal*, 40, 261–8.
Wilcox, R. R. (in press a). Detecting nonlinear associations plus comments on testing hypotheses about the correlation coefficient. *Journal of Educational and Behavioral Statistics*.
Wilcox, R. R. (in press b). Pairwise comparisons of trimmed means for two or more groups. *Psychometrika*.
Wu, C. F. J. (1986). Jackknife, bootstrap, and other resampling methods in regression analysis. *The Annals of Statistics*, 14, 1261–95.
Yuen, K. K. (1974). The two sample trimmed t for unequal population variances. *Biometrika*, 61, 165–70.

INDEX

Absolute error, 23–24
   and the median, 23
   and regression, 27
Alcoholism, 80, 102
ANOVA, 89
Asymptotically correct, 87

Bagplot, 200
Barnett, 135
Bayes, 5
Berger, 178
Bernoulli, 3
Bessel, 5
Bias
   and power, 80
   of Student's T test, 80, 85, 90
   see also hypothesis testing
Binomial, 3

Biweight, 151, 211
Boik, 236
Bootstrap, 80
   bootstrap sample, 96
   equal-tailed, 176
   general strategy, 95
   and heteroscedasticity, 107–108
   and least squares, 107–108
   modified percentile, 109
   and Pearson's correlation, 112–113
   percentile, 95–99
   percentile confidence interval, 98
   percentile t, 99–105
   percentile t versus percentile, 175–176
   and test of independence, 190
   and Theil-Sen, 208

Bootstrap *(continued)*
    and trimmed means, 168
    two-sample case for means, 105–106
    two-sample case for trimmed
        means, 170–172
    versus Student's T
Bootstrap t, *see* percentile t
Boscovich, 25, 27, 65, 107, 209
Box, 140, 158
Boxplot, 36–38
    finite sample breakdown point of, 38
    see also outliers
Bradley, 78
Breakdown point, *see* finite
    sample breakdown point

Cassini, 24
Central limit theorem
    and accuracy, 50
    and confidence intervals, 59
    derivation, 5
    and mean, 39
    and median, 44–45
    and outliers, 44
    and regression, 46–47
    and sample size, 5, 8, 39–44
Central tendency, 14
Cliff, 193, 203, 228, 233, 240, 243
Coefficient of determination, 186
Cohen, 127, 135
Confidence interval
    general formulation, 63
    for mean, 60–62
    for slope, 64–65
    see also heteroscedasticity
Correlation
    biweight midcorrelation, 193
    Kendall's tau, 191–193, 211, 228
    based on M-estimators, 193
    and outliers, 195
    Pearson's, *see* Pearson's correlation
    percentage bend, 193
    as a regression estimator, 210
    based on robust estimator, 224
    Spearman's 190–191
    Winsorized, 189–190

Covariance, 110, 198
Cressie, 87, 90, 03
Curvature, 183–184, 200–202
Dana, 82, 171
Davison, 115
de Moivre, 3
Dependence
    defined, 60
    between mean and variance, 75–77
    among ordered observations, 160–162
Deepest regression line, 223
Delaney, 237
Diabetes, 55, 129, 183
Diaconis, 158
Distributions
    contaminated, 117
    exponential, 41
    Laplace distribution, 13, 54
    lognormal, 76
    mixed normal, 117
    normal, 32–34
    uniform, 40
Doi, 130

Earleywine, 80, 102
Earth, 24–26
Effect size, 127, 134
Efron, 94, 115
Ellis, 24, 150
Equal variances, *see* homogeneity of
    variance
Equivalence, 176–178
Expected value, 2, 24, 51

Familywise error rate, 238–240, 243
Finite sample breakdown point, 17
    choice of in regression
    of least squares estimator, 29
    of MAD
    of the mean, 17
    of the median, 19
    of M-estimator, 154
    of Pearson's correlation, 113
    of Theil-Sen estimator, 208
    of the trimmed mean, 147
    of the weighted mean, 20
    of the variance, 21–22

INDEX

see also outliers
Fisher, 8, 68, 89, 235
Freedman, 158
Friedman test, 240
Frequentist approach, 58
FWE, see familywise error rate
Gasko, 118
Gauss, 4, 7, 24, 27, 50–52
  and least squares, 28
Gauss-Markov theorem, 56–58, 65
Generalized variance
  see variance
Glass, 91, 140, 158
Goldberg, 130, 135
Gosset, 74, 77, 80, 90
Guo, 178

Hampel, 117, 158, 178
Hastie, 135
Hawkins, 228
Heteroscedasticity, 56
  and ANOVA, 140
  and power, 86
  and Student's T, 85–87
  and Type I errors, 87
  effect on probability coverage, 65–66
  see also Pearson's correlation;
    percentile t
Hettmansperger, 234, 240, 243
Hinkley, 115
Hochberg, 243
Homoscedasticity, 56
Homogeneity of variance, 8
  and confidence intervals, 63–64
  see also homoscedasticity
Hsu, 178, 243
Huber, 117, 157, 158, 178
Huber's Ψ, 151
Hubert, 228
Hypothesis testing
  basics, 68–72
  and bias, 71

Iglewicz, 130, 135
Independence, 60
  test of, 189
Interquartile range, 37

Kendall, 234
Keselman, 90, 140, 158
Kruskal-Wallis test, 239

Laplace, 5–7, 12, 19, 24, 50–52, 58,
  62, 68, 74, 83, 90, 94, 107, 150
  distribution, 13
Least squares
  and the mean, 23
  as a measure of error, 22–24
  and normality, 5
  as a weighted mean, 28
  see also absolute error; least squares
    regression
Least squares regression
  computation, 28–29
  origin, 27–28
Least median of squares, see regression
Legendre, 27
Leroy, 135, 228
Lewis, 135
Leverage points, 130, 218
Liu, 200, 203
Location
  measures of, 14
  population, 14
  scale equivariant, 152
Long, 228
Luh, 178

MAD, 35
  and M-estimators, 152–153
  and outlier detection, 36
Mann-Whitney test, 231
Masking, 35, 134
Maxwell, 237
MCD, see minimum covariance
    determinant
McKean, 243
Mean, 14–17
  accuracy of, 50–55
  and asymmetry, 18
  dependence on variance, 75–77
  effects of outliers on, 16
  as expected value, 21
  lack of robustness, 149

Mean *(continued)*
  as an *M*-estimator, 151
  variance of, 60, 160
  Winsorized, 163
  see also weighted mean; median
Mean squared error, 50
Median, 17–19
  and asymmetry, 17–18, 127, 148
  and central limit theorem, 44–49
  and normality, 145–146
  as an M-estimator, 151
  population median, 17
  sample median, 17–18
  versus mean, 52–55, 123–124, 145–146
Median absolute deviation statistic, *see* MAD
Medina, 201
M-estimator
  adjusted, 221
  computing, 153–154
  defined, 153
  inferences about, 174–175
  one-step, 154
  origin, 24
  regression, 220–222
  adjusted *M*-estimator, 222
  versus trimmed mean, 154–157
Miller, 16
Minimum covariance determinant estimator, 197–200
Minimum volume ellipsoid estimator, 197
Multiple comparisons, 236–240
MVE, *see* minimum volume ellipsoid estimator

Naranjo, 240, 243
Newton, 24, 25
Neyman, 68
Neyman-Pearson, 8
Nonparametric, *see* rank-based methods
Normal curve
  basic properties, 32–33
  derivation, 3
  empirical checks of, 5–6
  equation for, 32
  Gauss's initial argument for, 4
  and least squares, 5
  naming of
  standard normal, 32
Null hypothesis, 68

Odd functions, 150
Olive, 228
Outliers
  and bivariate data, 134
  detection among multivariate data, 200
  detection assuming normality, 34, 126
  detection using a boxplot, 37
  detection using MAD and median, 36
  effect on Kendall's tau, 195
  effect on least squares regression, 131, 134
  effect on Pearson's correlation, 114, 194
  effect on Spearman's rho, 195
  effect on sample mean, 16
  and finite sample breakdown, 35
  and *M*-estimators, 153
  regression, 217
  see also power
Ozone, 230

Parelius, 200, 203
Pearson, E., 68
Pearson, K., 6
Pearson's correlation
  estimation of, 110, 112–113
  and heteroscedasticity, 111–112
  interpretation of, 179–183, 184–187
  and nonnormality, 132–133
  and outliers, 114, 194
  and slope, 180, 186
Peckham, 91, 158
Pedersen, 16
Percentile t, 99
Permutation test, 235–236
Population mean, *see* mean
Power
  defined, 69
  and heavy tails, 121–122, 134, 239

and outliers, 82, 172
and sample size, 70
and standard deviation, 71–72
and type I error, 70
Probability curves, 12–13
  heavy-tailed, 44
  light-tailed, 44
Probability density function, 12

Quartiles, 36
Quetelet, 6

Ramsey, 91
Random sample, 246
Rank-based methods, 229–235
  versus robust estimators, 233
Rasmussen, 91
Reading ability, 130
Redescending Ψ, 152
Regression
  based on ranks, 240
  choosing an estimator, 224
  comparison of methods, 226
  deepest regression line, 223
  and earth's shape, 26
  estimation, 26
  failure with a high breakdown
    point, 214–215
  L_1, 211–212
  least absolute value, 211–212
  least median of squares, 217
  least trimmed absolute value, 216
  least trimmed squares, 212–213
  M-estimator, 220–222
  multiple predictors, 27
  Winsorized, 210
Regression outliers, 130, 217–218
Relplot, 195–196
Residuals
  defined, 26
Richer, 24
Ronchetti, 158, 178
Rosenberger, 118
Rousseeuw, 135, 158, 197, 200,
  202, 203, 226, 227, 228
Ruts, 200, 203

Salaries, 221

Sanders, 91, 158
Scheffe, 89
Schizophrenia, 236–237
Schweppe weights, 221
Self-awareness, 61, 82–83, 171
Sen, 207
Sexual attitudes, 15–16
Shao, 117
Sheather, 127, 158, 178
Simon, 94
Simpson, 5
Singh, 200, 203
Smoother, 130–131, 201
Software, 241
Squared error, 22–24, 60
  see also least squares regression
Stahel, 158, 178
Standard deviation, 21
  see also power
Standard error, 59
Standardized difference, see effect size
Standardized variables, 186
Stars, 113–114
Staudte, 127, 158, 178
Stigler, 4
Stuart, 234
Student, 74
Student's T, 74–75
  expected value of, 80
  and inferences about trimmed
    mean, 166–167
  and nonnormality, 80–81, 85–87
  versus Wilcoxon test, 232
Summation notation, 245
Sunburst plot, 200
Symmetry
  and accuracy of the mean, 18–19
  assumption of, 5

Tamhane, 243
Theil, 207
Theil-Sen estimator, 208–210
  versus adjusted M-estimator, 222
Tibshirani, 115, 135
Tied values, 232
Transformations, 82, 90

Trimmed mean
  a common error when using, 160
  computing, 142
  confidence interval for, 167
  how much to trim, 143, 147
  and hypothesis testing, 166
  and nonnormality, 167
  and skewed distributions, 146–148
  two-sample case, 170–172
  variance of, 164, 170
  versus the mean, 143–145, 164–165, 167, 169, 171–173
  versus M-estimator, 154–157
T Test
  interpreting, 87
  one-sample, 72–78
  and nonnormality, 78–82, 85–86, 120–123
  two-sample, 84
  see also bias
Tu, 115
Tukey, 89, 117, 120, 135, 200, 203
Type I error, 69
Type II error, 69

Unbiased estimator, 211
Unbiased tests, 71

van Driesen, 197, 202, 226
van Zomeren, 203
Vargha, 241
Variance, 20–22
  effect on normal curve, 32
  estimation of, 21
  generalized, 197–198
  graphical interpretation, 124–125
  and heavy-tails, 120
  of mean, 246
  population defined, 246
  and power, 71–72
  and sample size, 5, 8, 39–44
  of a sum, 246
  Winsorized, 163
  see also masking

Weighted mean, 19–20
  and accuracy, 51
  and regression, 28
Welch's test, 171–172
Westfall, 243
Whiskers, 37
Whitford, 87, 90, 93
Wilcox, 91, 115, 135, 158, 175, 177, 200, 203, 228, 238–241
Wilcoxon test, 231
  versus Student's T, 232
Wilks, 197
Winsorized correlation, *see* correlation
Winsorized mean, *see* mean
Winsorized regression, 210
Winsorized variance, *see* variance
Winsorizing, 162–163, 187–188
Wu, 115

Young, 243
Yuen, 170, 178

Printed in the United States
134496LV00003B/62/A